Build Your Own
**Use what you
have to create
what you need**

Build Your Own

Use what you have to create what you need

Harrison Gardner

Gill Books

Gill Books
Hume Avenue
Park West
Dublin 12
www.gillbooks.ie

Gill Books is an imprint of M.H. Gill and Co.

978 07171 92649

Designed by Graham Thew
Copyedited by Jane Rogers
Proofread by Neil Burkey
Indexed by Eileen O'Neill
This book is typeset in 10.5 Sofia Pro Light.
Printed by L.E.G.O. SpA, Italy

For permission to reproduce photographs, the author and
publisher gratefully acknowledge the following:

© Ceresse Gardner: 197; © Erin McClure: 5, 8, 9, 19, 20, 74, 75, 76, 77, 79, 82, 83,
97, 149, 152; © Getty Images: 110; © Lorena Presno: x, 74, 127, 128; © Mahavir
Acharya: 6, 7, 8; © Unsplash: 149; © Wikimedia Commons: 201, 202, 203.

All other photography © Shantanu Starick

Essential tool illustrations on pages 15–18 © Erin McClure and Joseph Engstrom

All other illustrations © Erin McClure

FOR ERIN AND INARI,
BOTH MY GREATEST TEACHERS
AND MY BEST STUDENTS.

Harrison Gardner is an Australian eco-builder and sustainability designer based in the West of Ireland. Harrison is co-founder and executive director of Common Knowledge, a not-for-profit that aims to empower people with the skills and solutions for living a sustainable life. With over ten years' experience designing and managing the construction of conventional and off-grid buildings throughout Europe, Africa, Asia, the USA, Canada and South America, Harrison's focus is on constructing community spaces, teaching individuals how to build and maintain their own homes, and sharing core construction principles and techniques. He built his own home and has taught hundreds of people to do the same.

CONTENTS

Foreword

All of us have in the past imagined building our dream home, whether or not we remember it. Think back to those days and years spent building with wooden blocks and Lego bricks, creating idealised houses for ourselves and our animals, and drawing countless pictures in crayons and colouring pencils of the home we imagined we'd live in. We kept rearranging the blocks until everything looked and felt just right. In our drawings we made sure to show the building's orientation towards the sun, in the form of a shining yellow orb, and its relationship to its environment, often represented by a single tree, with birds in the sky, a meadow or lawn surrounding it, and occasionally hills in the background.

Shelter-making is an instinctive urge and we're as hardwired to do it as birds are to nest or spiders to spin their web. Yet nowadays, for many of us, the idea of daring to pick up a tool or engage directly with building materials brings up complex feelings that range from fear to exhilaration.

At our most aspirational, we feel a thrill of excitement at the notion that we can shape our surroundings. It's empowering and reassuring to know we can construct things that will make our lives easier, more enjoyable and more fulfilled – whether it's a garden shed, a pizza oven, a patio, a guest bedroom, or an entire house. But the fear inevitably arises in many of us. We feel unqualified and incapable. The consequences of getting it wrong are daunting.

Intuitively, we understand that our ancestors have always built their own shelters and other constructions to ease their lives, but the knowledge of how to do this has been lost over a few generations and we've become convinced that we've no alternative but to hire professionals to do it for us.

I experienced the anxiety myself when I set about building my own home in 1997. I had no experience whatsoever but had seen people building houses for themselves in Africa, India and South America, using whatever material lay around them: rocks, reeds, timber, slate, straw and mud, and I determined to do the same. The farmers around me in Westmeath baled oat and barley straw, and so I decided to use those as my blocks – 200 bales of them, laid like Lego bricks to form a small one-room structure on a thin band of concrete foundation and plastered with a mix of powdered white lime and sand, with a roof of corrugated sheeting.

I can't claim it was an easy experience, but it was phenomenally rewarding. I managed to create a little home for around €7000 that sheltered me over six years while I amassed the funds (and confidence) to construct, in 2002, a slightly larger two-room house of concrete blocks, plastered with mud, lime, straw, and with a grass roof. This second house cost me around €26,000 and I'm still living there today.

I would have given anything to have had a book like this one available to me at the time. There were simple 'how-to' books that showed how to install the electrics and plumbing, but there was never anything like *Build Your Own*. This is the one gift I'd love to be able to give to my frightened, uncertain self of 25 years ago. The answers to all the questions I had back then are contained within it, as well as to so many others that have come to light in the years since, as I've tried building compost bins, a glasshouse, a timber store, hen shed and tree house. In fact, had I read this book in 1997 I might still be living in that first straw bale house, as the reason it had to be knocked was that I had created inadequate foundations, not realising the principles behind siting a building on the land. (Also, it didn't have planning permission!)

In these pages Harrison Gardner lays out ways for us to be able to respond to our instinctive urge to create cosy, secure, personalised shelters for ourselves and our animals. He offers us an alternative to shackling ourselves to a crippling mortgage for much of our lives in return for a dehumanised, identikit box built with haste by the construction industry.

In some ways, it's a revolutionary book, challenging the many subtle ways that those in power lead us to believe that building is beyond our abilities and that we should enslave ourselves instead to a financial institution so that we can pay experts to do the work that our great-grandparents did themselves. It suits politicians and big corporations to have a society of biddable subjects whose freedom to redirect their lives, question the status quo, drop out of the workforce, or demand radical societal change is restricted by their obligations to banks and creditors. Not only does it ensure compliance and control, it also generates vast profits through taxes and interest payments.

If this book gives you the confidence to try laying a patio, or building a garden wall, it will have earned its keep; even if it just gives you the reassurance of knowing how your wastewater system works or how you could install a solar PV system. But, ultimately, it has within it the knowledge to empower you to shape and craft your own living space, and to alter and adapt the space as your needs change through life. It's a gift of empowerment in the form of print on paper and is in turn the fruit of all the gifts Harrison has received while building homes and public buildings throughout the world.

So, if you are ready to really take back control of the reins of your life, read on. Cast your fears aside – the revolution starts now.

Manchán Magan, 2021

Teaching alternative
construction techniques
in Argentina, 2017.

INTRODUCTION

Anyone can learn to build. And perhaps everyone should.

Whether you are an experienced building contractor or have never held a hammer, the same laws of physics apply to our structures. Being pulled in the various directions of affordability, practicality, passivity and longevity when planning a build can be vertigo-inducing enough to turn some people off the prospect altogether. But before you go down a spiral of concern and uncertainty, let me assure you that you can do this. You can understand the relevant concepts and materials, and how they interact. And, most important, you can ask, with confidence, why something should be built a certain way.

It is my hope that after reading this book you will begin to look at buildings differently, understanding how they were constructed and why they are still standing. Through studying the forces of nature at work, we can explore the values of different materials and construction methods and consider their alternatives. I want to equip you with the skills necessary to turn your building dreams into clear, tangible realities so that when you imagine the outlines of your future space, you will also be seeing the foundations, the wall structure, the cladding, the roofing and the finishes. I want to encourage you to indulge your imagination, but from an informed perspective, as it is here that I believe the instinctive builder in you needs to start. And all this is before you have even picked up a hammer! We will get to that shortly.

Now, I understand that you might be thinking that this book doesn't feel nearly heavy enough or have enough pages to carry a title like *Build*

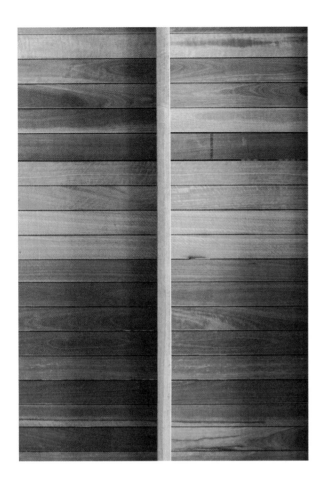

wall all experience, and resist, forces in the same way. I want to explain why steel is strong, and why most roofs are triangular. I want to decode some of the frustrating construction terminology which may have at times excluded you from the building conversation, such as U-values, airtightness and passive solar gain. Above all, I want to suggest a different way of achieving our basic need for shelter, which requires smaller loans, more time, and a softer, more harmonious footprint on our environment.

So why is this important? Why should you care about how your house works, as long as it does work? Why not just pay a 'professional' to do it for you? The reason, I believe, is instinctual. Food and shelter are our baseline physiological needs as humans. Whether we are chasing the skills of our ancestors or are forging new skills in the face of financial necessity, making, maintaining and mending our homes is part of what makes us human. Nourishing this area of our lives allows a naturally occurring aspect of ourselves to be seen and heard. You don't have to look far to see our cultural zeitgeist is beginning to feel disillusioned by the way our basic needs are being provided for us, ready wrapped and straight off the shelf. Can someone who has never met you truly understand your needs and design something ideal for you? A home ready for you and all your future needs?

Your Own: Use what you have to create what you need. Let me explain. This title may be the only way to really convey my feeling that the multitude of building practices, along with their various methods, all stem from the same basic principles. Understanding these principles will enable you to build with whatever materials and methods you choose, or have available to you.

Not only do I want to share with you the differences between certain materials; more important, I want to highlight their overlapping similarities. We can start by looking at how they interact with you, your home and with nature. I want to share with you why it is that an earthen wall, a brick wall and a timber frame

The concept of a 'finished house' is a relatively new one – two or three generations old, at most. Houses were never meant to be finished. They are utilitarian at their core, created to give us shelter and reflect our individual needs, cultures and lifestyles. They should grow and adapt with us, reflecting our ever-changing needs, as our homes of the past did. Bedrooms were added as children arrived, kitchens

opposite Exposed
framing with native
Australian jarrah.

left Solar-oriented
greenhouse in Co.
Clare.

were expanded, long conversations over glasses of wine led to cabins and conservatories. When the children grew up and fled the nest, bedrooms were transformed into art studios or reading rooms. These adaptations and changes were built by the owners, because building was part of life, no different from growing and cooking food.

Then, somewhere in the last hundred years, homes became a product which other people would create for us. Convenience, and the insurance of investment, became paramount, and we forgot how to build. We forgot how to think and talk about building, and we forgot to share the age-old tradition of building together. The good news is that it is never too late to awaken this instinct again.

We are instructed, and sometimes even forced, to build our homes with materials that are specified to meet building codes that have been defined by the insurance companies' engineers. They do this to protect the investment of the bank, from which most owner-builders will need to borrow money to pay the construction company, to build a home that meets the standards that have been set. This fear-fed cycle excludes anyone who lands too far away from the average. The best way to interrupt the cycle is to reduce the cost of building your home, and the first step to reducing costs is to take on as much of the work yourself as possible. Being willing to get your hands dirty will have the greatest impact on the overall cost of your project. This may seem like an intimidating notion at first, but so much of the building process is a series of repetitive, simple tasks. Moving earth, basic demolition, installing insulation, building internal walls, cladding and even plastering are all jobs that can be learned quickly and easily. Of course, there are some aspects of a

project that do require skilled professionals, but so much of it is actually very simple. Online tutorials can be really useful, and a professional you can call on for troubleshooting advice is an invaluable resource. Over the years, I have taught hundreds of novice construction students and I am yet to meet anyone who feels unable to participate on-site. I would really like to encourage you to feel the fear and build it anyway.

Of course, when it comes to using tools and safety on site, you need to develop some skills before diving in at the deep end, but this is something which can be learned from observation and practice. If only we grew up learning such basic survival skills! I often dream of a world where our core educational institutions taught every single student how to use these incredibly useful tools, empowering them with the skills to build their own homes. It seems illogical that we are not all taught how to use such tools, given that they are freely available to purchase over the counter and can pose a serious threat when misused.

Perhaps this sounds ideological, but just consider our driving education system. We have created a model which enforces people to learn how to safely drive any vehicle that weighs under 2000kg. Vehicle tests, insurance policies and road laws act in conjunction with the licensing system to produce limitless numbers of proficient drivers around the world every year. Could a similar system not be applied to every essential aspect of our lives, including building? One supported by a harmonious set of subsystems to ensure our legal, financial and insurance needs are met, just as we do with driving? If nothing else, such a model would certainly foster a culture of independent thinking and independent responsibility.

You may now be wondering what has led me to have such ideological hopes. I grew up in Australia and spent several years working on conventional construction sites there. Then I crossed the Indian Ocean to put those skills to work. In 2008 I landed in war-torn Western Kenya and launched into building schools and medical centres for impoverished communities. I learned more from my Kenyan building crew than I ever did working in Australia. Budgets, timelines and access to resources were at opposite ends of the spectrum in those two countries. There was something incredibly humbling about carrying all the water needed for the job site each day in 20-litre buckets up a slippery two-mile track. Every drop of water spilt was a personal torture, and of course wasting any water once back on site was

unheard of. We worked hard under the Kenyan sun, forever finding ways to stretch our budgets and make the process more efficient and affordable. In the communities I worked with, all members of the family were involved in the construction process. They were eager to learn and willing to help in any way they could, to save money and time. I built side by side with them as they taught me what hard graft truly was, and how different it felt to work for a purpose and not just a profit.

Some years later, I was riding through the Rann of Kutch in India on a Royal Enfield and I came across the Hunnarshala Foundation in Bhuj, not far from the border with Pakistan. I had stopped my motorbike on the dusty roadside to take a closer look at a small

lean-to structure that had caught my attention. The roof was shingled with car tyre treads, in a manner not dissimilar to the terracotta tiled roofs of France and Spain. As I was inspecting the structure, a man emerged from another interesting building. This one was constructed of red brick, clay mortar and earthen plaster, with a traditional thatch roof. Sandeep Virmani introduced himself as the director of the Hunnarshala Foundation and generously invited me in for an in-depth tour of their campus.

Hunnarshala focused primarily on creating recycled, sustainable construction products which were sold to high-end architecture firms in Ahmedabad, Delhi and Mumbai. Their products were made from industry waste and locally sourced natural materials. They sourced literal truck-loads of mixed waste construction material from Alang, the world's

largest ship graveyard, a few hours away. They ran an apprenticeship programme with the local youth community, who would trim, plane, glue and screw the materials back together into laminated boards, wall cladding and floors. I spent several weeks with Hunnarshala drinking chai, listening to stories and turning trash into building materials. We made unfired compressed earthen blocks by mixing clay, fine sand and lime. We soaked hessian sacks in clay slip and made roofing tiles by shaping them over our thighs. Alang's team taught me the importance of understanding the properties of a material, rather than only considering it for its prescribed use. I soon came to see a car tyre, not only as a combination of rubber and steel, but as an impressively strong and highly useful resource, one which was in fact over-engineered for many potential uses in our buildings.

right Vaulted
brick ceiling at
the Hunnarshala
Foundation campus in
Bhuj, Gujarat, India.

below right Earthship
tyre wall with
compressed earth in
Taos, New Mexico.

opposite Teaching
students about thermal
flow at one of our early
building courses in
Ireland in 2019.

Shortly after my time in India, Mike Reynolds of Earthship Biotecture welcomed me into his 'Garbage Warrior' fold – a travelling circus of unconventional builders who specialised in using a blend of natural, recycled and conventional materials to create autonomous, off-grid buildings. The job brought me to the far reaches of the planet as a site foreman. We built in some of the most remote and inaccessible regions of the world, from the mountainous heights of the Himalayas to the southern tip of Argentina. From the dry deserts of New Mexico to remote, balmy Pacific islands.

Mike's passion for creating living homes that are an extension of their inhabitants and are quite literally full of life, has surely left its mark on me. The word 'genius' comes to mind when I think of his methodical madness, bringing watercolour sketches to life as

buildings, and perpetually challenging the limits of materials and concepts, and even challenging the building crew to be the best possible versions of ourselves. It was during my time with Earthship Biotecture that I was given the opportunity to formally teach building and alternative construction practices. Sharing and discussing sustainable construction concepts such as rammed earth buildings, passive solar gain and rainwater catchment was immediately gratifying. I realised that I could be much more effective in the long term by focusing my energy on enabling people to build themselves, as opposed to just building for other people. This realisation propelled me further into education and led to the opening of our own small building school in the West of Ireland, where I now live.

For years, we building enthusiasts have met up around the world, stamped clay with our feet, turned piles of timber into buildings and discussed the intricacies of thermal movement and the role of gravity in construction. I want to welcome you with open arms to the conversation. Don't worry, you haven't missed much, we are really only just getting started.

The language of structure

Language is one of our greatest tools and simultaneously one of our most effective barriers. When we speak a language we are welcomed into a world of learning, education and possibility. When we are unable to communicate, we can end up having more questions than answers. Naturally, as we go deeper into our subject matter, the language becomes more complex. It can be difficult to unlearn such complexities once we understand them and professionals in the construction industry often appear to be speaking an entirely different language. I would like to encourage you to persevere through what can often feel like a communication challenge. Just as when you travel to a foreign country and smile and bumble your way through many interactions, sometimes receiving food you did not order or offending people you were trying to compliment, you will, if you are new to this, likely experience similarly confusing interactions in the foreign land of construction and material suppliers.

above Pine cut and dried into dimensional lumber, ready for construction.

opposite Builders' providers in rural Co. Clare.

How to use this book

I do not believe there is a one size fits all house design. The way it looks and functions should be informed by your personal and cultural needs, as well as responding to the realities of your site location and climate. Within one book, we cannot possibly cover every single building technique and discuss the fundamentals of every single material – especially given my belief that most things on planet earth could be used as a building material, in some way or another!

Instead, it is my aim to help you make informed, contextual design and construction decisions and have the confidence to get involved, whether it is just in discussion with your contractor or by getting your hands dirty and building yourself. Each chapter within this book focuses on a key benchmark of the building process, starting with foundations, and working our way up and outwards to create a finished structure. We will explore each topic and look at the most commonly used material options. This is followed by a series of practical tutorials which illustrate the methods and practices discussed.

The recipes within the practical methods may need to be closely followed when you start out, but once you begin to understand what each ingredient is actually offering your overall mix, you will be able to adapt the recipe to suit your project – whether it is a little more sand in this plaster mix, or an extra nail in this connection. Construction is not something you dabble in once, then never pick up again; the knowledge you gain from being actively involved on-site will remain with you.

I hope that, by reading this book, you feel liberated from the idea that there is any one 'right way' to build a structure. I want to encourage you to improvise, to think outside the box, and be creative. Allow yourself room and time to experiment and make mistakes. Looking for definitive answers can be a barrier to understanding. It is precisely our awareness of our naivety that makes humans so incredibly efficient at learning and evolving. And whatever you are doing, whether it is building related or not, be sure it brings you joy.

opposite Outdoor breeze-through partition wall in Australia.

this page Steel-frame structure built using minimal materials organised into strong shapes and elevated from the ground.

Essential tools

Below are the tools which I consider to be essential when taking on a self-build project. Builders use their tools to earn money, and owning and learning to use your own tools is going to save you money too.

There is huge variety in the quality and price of tools, and it is often the case that the more you spend, the better the quality, or the longer the tool's warranty. Spend what you can afford, and take care of your investment.

TAPE MEASURE

Used to measure everything. Get a tape measure that is at least 8 metres long. If you are choosing from a range of tape measures, pick the most lightweight option, so that it can be on your person at all times. Whether clipped to your tool belt or in your back pocket, your tape measure is your best friend on a building site.

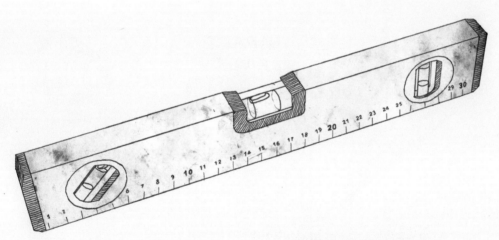

SPIRIT LEVEL

A spirit level is used to determine how level (horizontal) or plumb (vertical) something is. It is also a convenient straight edge which can be used to mark out lines.

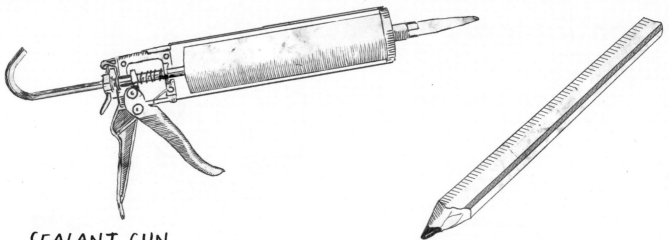

SEALANT GUN

Also known as a silicone gun, these dispensers are designed to suit a variety of adhesives and sealants in corresponding tubes.

CARPENTER'S PENCIL

It is no coincidence that most builders use a square pencil. They are designed specifically for use on a building site. The edges are flattened to stop the pencil rolling away when placed on a sloping surface. The lead is usually hardened, which makes marking concrete and stone easy.

MITRE SQUARE

This is a small, hand-held tool with a 90° corner and two 45° corners. It can serve as a protractor, a mitre square and a roofing square, all used to help us determine angles on our site.

IMPACT DRIVER

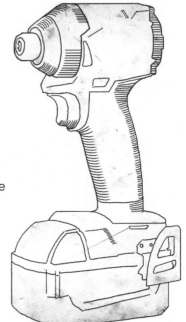

An impact driver is similar to a drill, with an added built-in hammer action. This makes it much easier to drive screws and fixings into timber and allows the batteries to last longer than a drill. I use an 18 volt cordless driver.

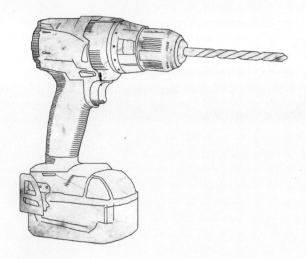

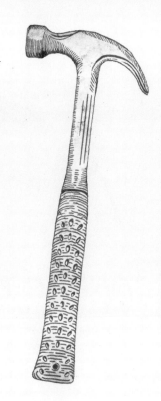

FRAMING HAMMER

An extension of your arm, your hammer may quickly become your most beloved possession. I have had a 22 ounce Estwing framing hammer for over five years, and it has built dozens of buildings with me all over the world. The balance, length and weight all affect how a hammer feels to swing and strike. Look for a long-handled claw hammer with a medium-weight head.

DRILL

Your drill is the tool you will use to drill holes. You can get a variety of drill bits in different sizes for different materials. You can also get various attachments, such as a whisk, for mixing plaster and paints. I use an 18 volt cordless drill.

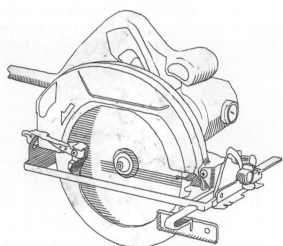

CIRCULAR SAW

Also known as a skill saw, the circular saw is a versatile cutting tool with an adjustable blade depth and angle. It can be used in conjunction with your mitre square to cut any angle. They are typically only used with wood. I use a 230 volt corded saw with a 175mm blade.

CHALK LINE TOOL

This is a retractable string, coated in chalk, used to mark long, straight lines. It works by snapping the taut line across your work piece and marking the surface with the chalk.

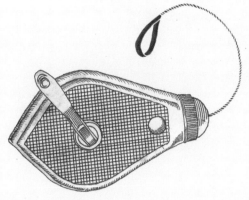

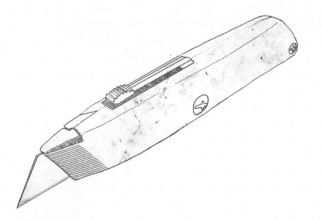

BLADE

This is a small, retractable knife with replaceable blades for cutting membranes, plastics and, most important, sharpening your pencil.

ANGLE GRINDER

We use grinders primarily for cutting and shaping metal. They are also useful for cutting off old screws and nails from recycled wood. They can be used with a range of blades for various tasks and materials. I use a 100mm, 18V cordless grinder as well as a corded 230V grinder.

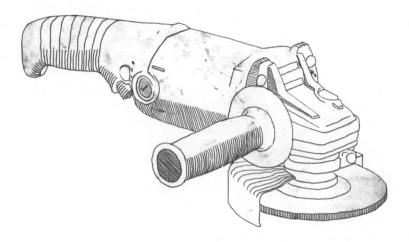

SAFETY EQUIPMENT

Every tool and process carries different risks and requires different safety equipment to ensure you can continue to build for many years to come. The basics you will need on every construction site are: ear protection, eye protection, dust mask, gloves and steel toe-capped shoes. Always take time before beginning a process to assess the potential risks to yourself and the dangers in the environment before starting. Take necessary precautions and, more than anything, use your common sense.

Earthship veterinary
clinic in Zuni Pueblo,
New Mexico, 2017.

CHAPTER 1

THE WEIGHT OF GRAVITY: THE PHYSICS OF FOUNDATIONS

left Inspecting a stone wall with Inari before carrying out repairs.

below Dry-stacked stone wall in Co. Clare.

opposite Brickwork beneath plaster and paint, Australia.

Between the ages of 12 and 17, I managed to break 13 bones in my teenage body, all on different occasions. This wasn't a result of inept parenting but rather my own bumpy journey to discovering that no matter which way I rolled the dice, I could not win a wrestle with gravity. I now watch my ten-month-old daughter learning the same lesson, as she heaves herself from her back to sitting up, only to tip over sideways. Naturally, she finds this impossibly frustrating. My now healed bones echo with empathy for her, as the force of gravity is a relentless bully. Luckily for us, it is a predictable one.

We experience the pull of gravity in a similar way at any given location on Earth, towards the ground, or, more precisely, towards the centre of our planet. Understanding this natural phenomenon allows us to build impressively tall buildings with the confidence that they will stay up. When it comes to smaller, more humble structures, the same laws of physics apply. They also apply to the structure of our bodies. This is perhaps not so surprising, when you consider that we are all, on an atomic level, made up of the same stuff. All matter has a certain amount of gravitational pull, proportionate to its mass. The greater the mass, the greater the gravitational pull, which is why we are pulled towards the earth instead of being pulled towards each other. Although there are counter-forces at play, the planet, due to its sheer size, always wins.

Exercise
OUR BODIES AS BUILDINGS

Our bodies are probably the easiest place to understand the effects of gravity, so I want to encourage you to get active for a moment. Go outside and find some solid ground to stand on, whether it is grass, soil or sand. Stand with your feet together and try to feel the gravitational pull on your body. The mass of your body is being pulled towards the ground and your feet are providing the resistance. You can imagine that, without our wide, flat feet, our legs might penetrate straight into the ground, much like a cow's hooves in mud. A smaller surface area provides less resistance to the force of gravity on your mass. It is no coincidence that in the language of construction, we refer to foundations as footings. Next, spread your legs shoulder-width apart to create an arch. This arch supports a tower, with two posts supported by two footings. While you are standing in this position, consider the elements of your body that are making this position possible. The combination of your rigid bones and elastic muscles are working together to resist the pull of gravity.

On a subconscious level, we are all aware of the unseen force of gravity. We instinctively make allowances for it, as well as placing great expectations on its presence. Right from the construction of our first prehistoric structures, architecture and engineering can be viewed as a response to the perpetual phenomenon of gravity and an answer to the question, How do we stack and combine materials in such a way to resist and redirect the force of gravity?

Foundations: redirecting the force of gravity

Now that we have a basic idea of the effects of gravity on our structures, we can look at the pivotal role of foundations. The average three-bedroom home weighs somewhere between 36 and 80 tonnes, which is the equivalent of 6 to 12 adult elephants, or 500–1000 people. Anyone who has been at a rainy music festival has seen what the weight of 1000 people can do to a landscape. The perpetual pull of gravity, which pushes the weight of our house down into the ground, requires an equal and opposite resistance to prevent it sinking down into the Earth's surface.

In the past, before we had the skills and tools to create concrete, natural bedrock and rocky outcrops were sought after as solid foundations for society's more substantial structures – castles, forts and towers. Even the Egyptian pyramids, which appear to be floating on a sandy sea, are in fact built on the huge layer of limestone bedrock that lies beneath the desert sands. Let's use our body again to help us to understand why our ancestors decided to build on foundations of stone, as opposed to sand or earth. Imagine walking barefoot on a smooth, sandy beach. We can feel the ground compress beneath our feet as we walk, or even be pushed out to the sides and up between our toes. Compare this to the sensation of solid resistance when walking on stone or concrete, and we can start to understand the strong inclination to use solid foundations as a means of connecting the massive accumulative weight of ourselves, our buildings and their contents to the Earth.

Finding exposed stone was not always easy, and having the resources to dig down to the bedrock was not feasible for most, so the average medieval owner-builder had to be a little more creative in their methods. When digging down to undisturbed

earth, which is earth that has not been tilled or broken apart in recent history, they found that the naturally compressed clay and sand was, in fact, stable enough to function as a base for the foundations of their homes. This undisturbed earth lies at different levels depending on where you are, but is most often found within a metre or two of the surface. Stones were used to fill the trench and raise the building up out of the ground.

Certain methods were more successful than others. The best results came from buildings where the stones were stacked in such a way that they would not shift or split under the load. And so we came to understand that the role of foundations was not only to connect us to solid ground, but to evenly disperse the weight of the building over as great an area as was necessary. Those medieval three-dimensional jigsaw masters had a clear, instinctual

understanding of what is now known as the load path and how best to design for it. They discovered that if they stacked the stones in even layers, with overlapping joints, their heavy stone walls could not only be self-supporting, which was their primary function, but could also carry large amounts of additional weight. This method of making foundations was implemented for thousands of years, and fine-tuned as we spread across the land building structure after structure, digging down to undisturbed earth and building our way back up.

In areas on or near to water, it wasn't quite so simple. Whether it is the ocean, a river or a marshland, the presence of water in soil makes it much more difficult to find solid ground. But necessity has always been the mother of invention, and being near a clean water source was also necessary for survival. And so we adapted again. Wooden post foundations have

been found in Switzerland dating back as far back as 12,000 years ago. Driven into the soft earth of shallow lakes, these foundations kept the dwellings and the inhabitants out of reach of predatory animals and flooding. The same techniques were used to expand cities situated near, or on top of, the water sources they required to function, with the wooden posts needing constant repair as they deteriorated in the shallow waters. Experimenting with different types of wood, as well as the use of preservatives and treatments, continues into the present day. Since the 17th century, Boston city, Massachusetts has been expanding across its waterways, adding over 5000 square kilometres of man-made land atop deep-reaching foundation pylons, which penetrate the subterranean clay layer.

The Romans were the first humans to refine the use of concrete, but it wasn't until we started adding steel to the mix in the 1850s that we created what we now know, love, use and abuse as reinforced concrete. Concrete's rise to popularity is unlike any other evolution we have seen in the construction of our houses. No longer limited by the materials that we could cut or carve, we could now create our very own man-made stone, tailored to varying strengths and shapes, according to our needs. Another reason for its immense popularity was its portability, as it meant that we no longer had to carry giant monolithic stones around. Concrete could be transported in small amounts to create large foundations, with huge load distributing capabilities. Add this to its relatively low cost and you can see how our stone stacked foundations were quickly regarded as obsolete. Mines opened up across the world to harvest the lime, sand, clay and stone we would need to ride the getaway horse known as concrete.

Minimal concrete foundations with structural steel, with very little ground disturbance, Australia.

Loads explained

Load can be defined as the weight or source of pressure which is experienced by our structures. Some load is referred to as dead, or static, load; and some is referred to as live, or active, load. Dead load does not move. An example of this would be the weight of your appliances, furniture and the building materials themselves. Live loads refer to the weight of occupancy, and suggest vibrations and movement. An example of this is the second floor of a building, where human activity would affect the load at any given time. Foundations are designed to work specifically with your building and to spread a point load, which is a load focused on any one point of the structure, right across the foundation.

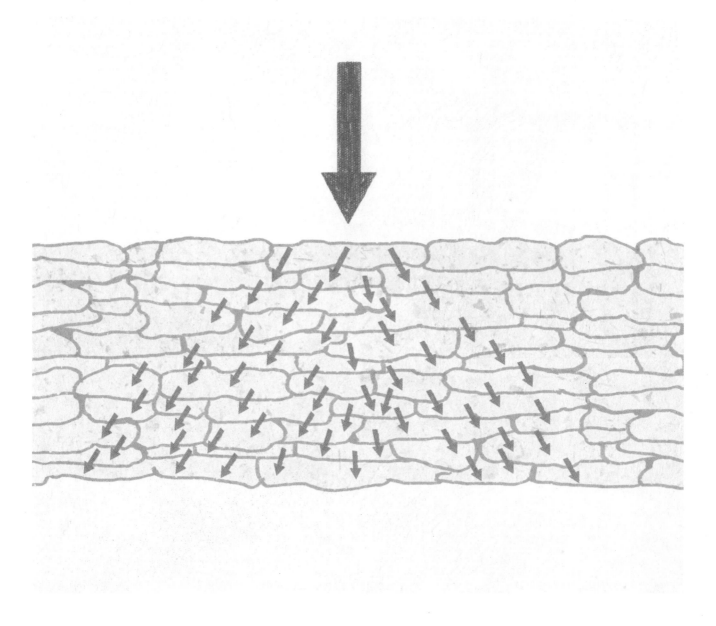

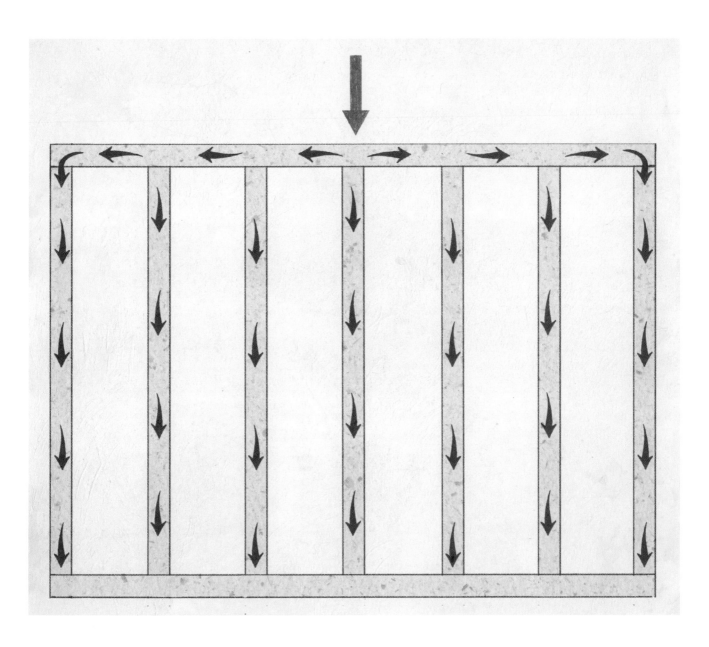

opposite Dry stack
stone wall with arrows
illustrating the load
path.

below Load path
through a conventional
timber stud wall.

Accurate site layout

When it comes to marking out the site for your foundations, ensuring the corner angles are correct is very important. Being just 1° off an angle will likely affect many other wall lengths and angles around the building. Luckily, there is a solution to this problem, which most of us will have been introduced to in school mathematics. Now we finally have a practical use for the mouthful that is Pythagoras's theorem!

Pythagoras's theorem is a mathematical law that tells us that the square of the hypotenuse of a right-angled triangle is equal to the sum of the two shorter side squares. This can also be expressed as follows:

$$a^2 + b^2 = c^2$$

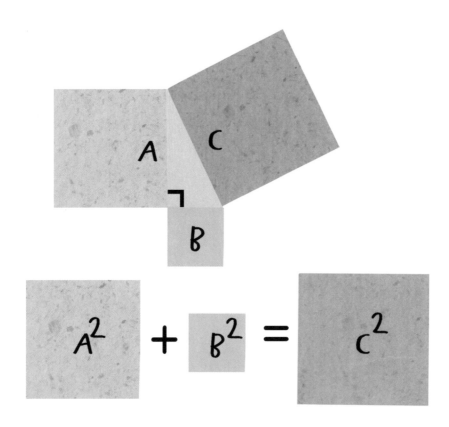

What this law inversely tells us is that if you can make sure that c does equal a + b, then you have yourself a right angle. To help with our understanding, there is a shortcut to this law. Conveniently, 3 squared (3 × 3) plus 4 squared (4 × 4) equals 5 squared (5 × 5). Which can also be expressed as 9 + 16 = 25. To visualise this, draw a right-angled triangle with a 3cm side and a 4cm side. You will find that the distance between those two points is precisely 5cm.

Now apply this to our real-world building site. If we measure 3 metres in one direction, and 4 metres in another direction, when the distance between those two points is exactly 5 metres, we know we have created a perfectly accurate right angle with lines that can be infinitely projected in those trajectories, maintaining their perpendicular angle.

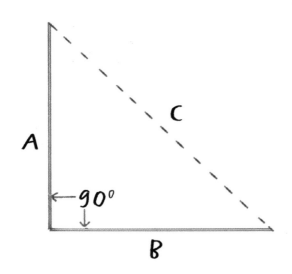

$$A^2 + B^2 = C^2$$

$$EG\ 3^2 + 4^2 = 25$$
$$9 + 16 = \sqrt{25}$$
$$= 5$$
$$\Rightarrow C = 5$$

Common foundation types

Let us now look at some of the most commonly used conventional and alternative foundation options for our buildings. While the materials differ, the same rules of physics apply to all the following foundation types. Their job is to safely carry and distribute the potential load of the structure down to the ground.

Isolated footings, also known as pad foundations, are concrete foundations that are placed in strategic locations to bear the weight of a building. They use carefully designed steel reinforcement to distribute the building's potential load. They are commonly used in steel frame construction, where the weight of the whole structure is split over a certain number of posts, which carry the weight to the ground. If designed to carry a substantial amount of weight, isolated footings will require a higher-strength concrete. The concrete mix and pad foundation size is typically specified by an engineer, according to the weight of the building it needs to support.

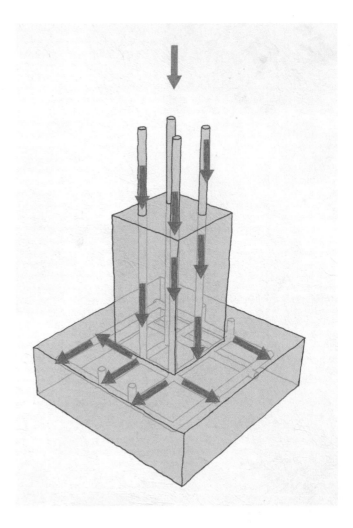

right Isolated footing with arrows illustrating the load path.

STRIP FOUNDATIONS

Strip foundations are linear foundations that travel under all the load-bearing elements of a building. In houses, this is usually the external walls and any load-bearing internal walls. Strip foundations typically have long lengths of strategically placed reinforcement steel embedded within the concrete to help disperse the potential load.

COMBINED FOOTINGS

A combination of isolated footings and strip foundations are often used together. They allow for precise and minimal concrete usage, whilst providing support exactly where it is needed.

RAFT FOUNDATIONS

A raft foundation is a slab of concrete which is poured over the footprint of the entire building it supports. Raft foundations are typically made with steel reinforcement bars, often inserted as a welded grid called wire mesh, to distribute the potential load. These are often the fastest type of foundations to construct, but use excessive amounts of concrete and steel, unnecessarily reinforcing areas that will not be carrying any load.

STONE FOUNDATIONS

While it is currently difficult to get planning permission for the use of stone foundations in Ireland, this foundation type is seen in older dwellings, and in retaining walls (i.e. walls that hold back or support soil laterally) around motorways and coastlines. Gabion cages are electro-welded or woven steel cages designed to hold the stones in place. We use them to build very high retaining walls, using the force of gravity to our advantage, by stepping them into the landscape.

TYRE FOUNDATIONS

Many countries, including France, the UK, the USA, Japan and Australia, have approved the use of stone-filled recycled car tyres as a foundation system. This method, pioneered by Mike Reynolds of Earthship Biotecture, uses the tyres to hold the stone in place, much like a gabion cage. Constructing layers of tyres with staggered connections ensures that the potential load is dispersed evenly.

RAMMED EARTH FOUNDATIONS

Rammed earth foundations are an excellent option in dryer climates. The process involves constructing a reusable formwork of timber and steel, into which a mix of clay, sand and occasionally cement or lime powder is poured. The loose mixture is then tamped and compacted, around 30cm at a time, usually by machine. Once it is compact, the formwork is removed and slid up the wall, so that the next layer can be added and compacted. This type of formwork is also used in concrete construction and is known as slip forming.

EARTHBAG FOUNDATIONS

Often referred to as super adobe, this natural foundation method is also approved in many countries, and used with great success. Earthbag structures and foundations involve filling polypropylene sacks, or a continuous polypropylene tube, with clay, sand and binder (cement or lime), similar to the rammed earth method. The bags are layered on top of one another, with barbed wire pieced between each layer to secure them in place during construction. Once complete, these walls can be plastered to make them weather- and UV-resistant.

Site selection: considerations when choosing land

The land you choose to invest your time, money and energy into is as much your home as the structure you build on it. All the geographical attributes of the site will transfer to your buildings, and should inform the way you approach your build.

Not many people find themselves exploring the world for their ideal site location. In most cases, we build on land which we can afford, or have inherited, or because it has pre-existing structures on it, which allow us to get started without the need for planning permission. If you do happen to be looking for new land to build on, there are several factors to consider. Working out what feels right for you may take some brainstorming, searching, compromise and gut-following.

opposite Looking southwest from our home in Co. Clare.

this page Exploring our land to understand where the sun reaches and where the water flows.

Exercise
HOME IS WHERE THE HEART IS

We all have a different idea of what our dream home looks like. We may be inspired by the houses we grew up in, or the spaces we have appreciated while travelling the world. We are all individuals, with different backgrounds and histories in relation to buildings. When you create a home with another person, your paths intersect and it can sometimes lead to conflicts in decision-making. When we are building a nest, it is normal for people to feel passionate, and tied to our ideals. If you find yourself in a conflict of ideas, start by investigating the experiences that may be informing your own design decisions.

The following exercise can help us to understand our hopes for the path forwards into our combined future and it works best when you do the exercise with the people you are going to be living with. Take your time to consider the things you spend the most time doing in your home. Use it as an opportunity to listen to yourself as much as an opportunity to communicate your ideas to someone else. On ten small pieces of paper, jot down the ten most important things that you hope your site and future home will provide for you. Some examples might be:

• Natural light
• Indoor plants
• Room for a vegetable garden
• Proximity to the ocean or a lake
• Projector screen for movies
• Large kitchen/bedroom/bathroom
• Space for animals

Resist discussing your choices until the end of the exercise. Remember that there are no right or wrong answers. Once you have written down your top ten priorities, try sorting them in order of most important to least important. This can be a surprisingly challenging task but will be highly valuable for your own decision-making in the future and the compromises you may have to make.

Now share and discuss your choices with each other. Explain each item in detail and enjoy the discussions around your similarities and differences. Remember to listen and be open to learning. You are not bound to your choices and the purpose of this exercise is to simply get the ideas flowing and help you identify things that are most important to you. You might surprise yourself!

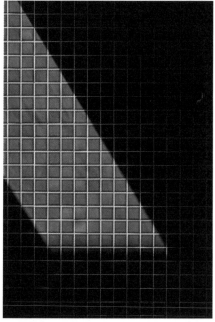

left Strong shapes celebrating the process.

top Solar (and thus lunar) oriented shower space.

above North-facing ash forest, Co. Clare.

Positioning your structure

Once you have determined the land you wish to work on, positioning your structure is the next most important decision. If you plan to renovate a pre-existing structure, you will want to capitalise on the site's main assets.

SLOPE

There is no such thing as flat land. The rain will fall and the water will run somewhere. Make sure you understand the topography of your land and where the rainwater runs and collects, so that you can build accordingly. The very basic rule here is that water runs downhill, so the closer you are to the top of a slope, the less water you will need to deal with. Ideally, a slight slope towards the Equator will give you maximum solar exposure and ensure that potential obstructions in front of you, like trees or other houses, will be lower down the slope than you, so they will never cast a shadow.

PERCOLATION AND SOIL SUITABILITY

Ensuring that the ground will absorb and filter your wastewater at a suitable rate is a necessary factor when applying for planning permission, if you are building in a rural environment. If you are in an urban environment, you may be able to direct your wastewater to the public drainage system. While investigating the ground conditions on your site, it is also interesting to see what the land is made up of, for both growing and building purposes. If you can gain access, I strongly suggest you dig a hole, approximately one metre across and a metre deep, somewhere close to your ideal site, to determine the make-up of your soil and sub-terrain. If you encounter water in this test hole, it is likely there is a high water table on your site, which may affect your foundation design. You never know what you will find. A high clay content soil will hold water, while a high sand content soil will drain it quickly. When digging for our own foundations we uncovered several tonnes of almost pure clay, which we have used over the years to make lots of free, natural paints and plasters.

COMMUNITY

If you are looking at land that is close to other houses, you may need to seek permission from the neighbouring community to build on it, so getting to know your neighbours can be useful – and nice – when you are new to the area.

SOLAR EXPOSURE

In the northern hemisphere, the passive and free heat of the sun comes from the south. Stand on the site where you imagine your structure to be and work out the path of the sun. Consider what elements of the landscape might interrupt your solar exposure, now and at the opposite time of year. Whether they are buildings, trees or hills, it is vitally important to relate your structure to our greatest source of free heat and light. You can learn more about understanding the path of the sun on page 166.

VEHICLE ACCESS

After years of hauling materials by hand (or mule) to inaccessible sites, I would strongly recommend choosing to build in a location that can be reached by vehicles. Particularly if you are taking on a large building project, where large trucks may be required for delivery of materials.

RECIPES

SOIL TEST

To use the soil around us to build with, we must first determine and correct its ratio of binder to aggregate so that we can trust it to disperse load well. A simple soil test is used to determine the composition of the soil on your land. Soil, also known as earth, is made up of clay, sand, minerals and organic matter such as decomposing tree roots and grasses. The useful parts for our building work are the clay and sand, and it is important to know in what proportions they are occurring on your land.

Ingredients
- A shovel
- A large glass jar
- Water
- Optional – pickaxe (if your ground is very hard)

Method

1 Find somewhere on your site to dig a hole. Try to find a place where you will be happy to dig more holes, because if you find that the earth is very suitable for cob you may need a lot more of it! Dig down past the topsoil and organic material until you find the sand, clay and mineral layer. Clay comes in many shapes and sizes, so do not give up hope if you do not find a big sticky ball right where you dig. It is often dispersed throughout the soil and can be crystallised, which makes it appear a lot like sand.

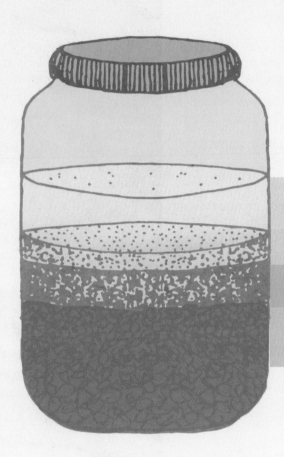

WATER

CLAY

SILT

SAND

2 You want to dig up enough of this material so that you can fill your jar ¾ full. Remove any big stones or grass roots.

3 Fill the rest of the jar with water. Now shake, shake, shake!

4 As you shake the jar, tiny clay particles will be released from the sand and stone they have been gripping to, and float into the liquid.

5 Place the jar on a level surface where it can be left, undisturbed, for several days. It takes just a few minutes for the stones to fall to the bottom and for the sand to rest on top of this. The clay will slowly settle on top of the sand over the next few hours, but it can sometimes take days to fully settle, as it is a very fine material, and easily floats into the water. The organic grasses and roots should float to the top of the clear liquid.

6 Once the layers have fully settled you can clearly observe the sand-to-clay ratio of your soil. Now to start building!

GABION CAGES

Gabion cages are steel cages that can be filled with stones and stacked like large blocks to form strip foundations. They distribute our building loads evenly and widely into the ground. They are a modern, engineer-approved, adaptation of dry stone stacking. Commonly used as retaining walls along motorways and by the seaside, gabion cages reduce transportation costs, as they can be filled with local material. They are not susceptible to water damage and can be assembled and constructed in remote locations where machinery cannot reach. In the domestic setting, gabion cages provide us with a great solution for retaining walls. You can order pre-cut, partially assembled gabion cages from a supplier, or you can use the instructions below to cut and make your own cages.

Makes a single gabion cage – 1m deep × 1m wide × 2m long

Ingredients

– 2 sheets of 50mm × 50mm × 4mm galvanised steel mesh. These usually come in 2.4m × 1.2m sizes
– Bolt cutters or an angle grinder
– End cutters
– Tie wire
– Stones 100–200mm diameter
– Optional – hog ring pliers

Safety note: Always wear gloves and eye protection when cutting and working with raw edged steel.

Method

1 First you need to cut your mesh using either bolt cutters or an angle grinder. You need six flat pieces of mesh to construct a three-dimensional rectangle: four pieces at 1 metre by 0.5 metres and two pieces at 0.5 metres by 0.5 metres. When cutting, ensure that you always leave a steel edge around the outside of your rectangle, as you will need this edge to tie your wire to.

2 Next, lay your four 1-metre pieces on the ground, with their long edges touching. Use your tie wire and end cutters to connect the mesh pieces together, starting at the corners. Loop the wire around the two edges you are connecting and make a single twist with your hand. Then use the end cutters to grip the twist in the middle. Continue to twist the wire for three more rotations, pulling away slightly as you twist. Place a tie every 0.5 metres along the edge. Join all four 1-metre pieces together on their long sides and then join the two 0.5-metre pieces on either side of one of the long pieces.

3 Lift up the sides of your cage and add the necessary ties to hold it in a three-dimensional cuboid shape.

4 Next, you need to prepare the ground where you will place your cages. Because they are made of mesh, which is slightly flexible, it is important that they are installed on stable ground. Remove any loose or unstable material. If the ground is particularly uneven, add 50mm of sand or gravel across the whole area to create a compact level surface. Place your cage or cages in their final position.

5 When they are in position, use tie wire to lace together the exposed edges of the cages. The goal here is to create a mesh formwork that is securely connected throughout. You can weave tie wire through the edges of the cage to connect multiple cages together. Use your intuition; if you think something doesn't look tight enough, add another tie. If you are making a lot of cages you might want to invest in a hog ring pliers, which is specifically designed for connecting mesh together.

6 Once your cages are securely attached together, they can be filled by hand or a machine, depending on the scale of your project. Choose stones which are strong and durable, ideally between 100mm and 200mm diameter in size. Take care not to damage or bend the cages as you are filling them, and be sure to fill all adjoining cages at the same time.

7 Once your basket is half full you need to brace the faces of the basket together. All exposed cage fronts should be braced. This can be done by weaving your tie wire from the front face to the rear face of the basket. Pull the wire tight and use a stone or a stick as a lever to tighten the wire, wrapping the tie wire around the cage at each face to secure it. Continue filling your basket to the top.

8 Baskets can be filled to approximately 50mm above the top of the cage to allow for settlement of stones. Tie the corners of the lid down first and then tie and weave the rest of the lid in place.

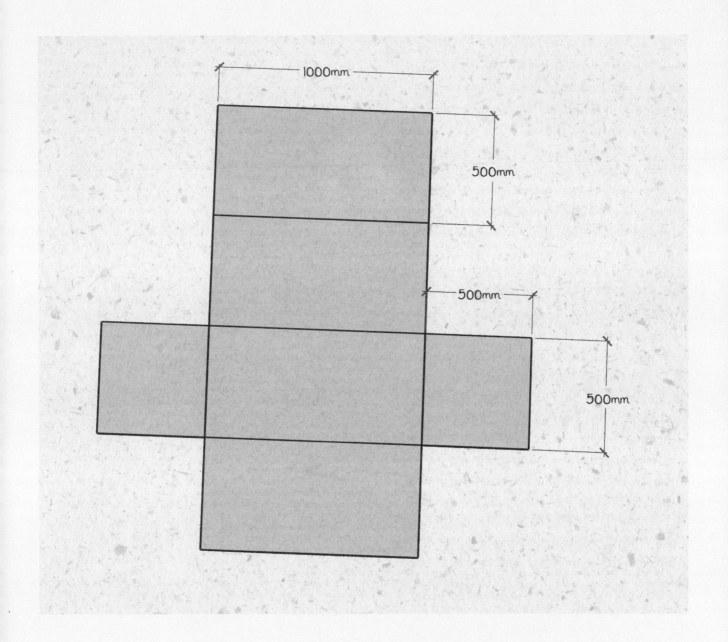

STONE STACKING

There are certain building techniques that seem just too simple to build a home with. Yet we are surrounded by exceptionally well-built stone homes, many of them over a century old. The traditional Irish cottage is typically made with solid walls constructed of stones stacked one on top of another. The gaps between the individual stones are filled with small stones and a mud or cob mixture known as loose fill to form a solid wall. Long through-stones are also used to tie the inner and outer sides of the wall together. These walls use the most simple form of load division and distribution, and we are repeatedly challenged to find the right stone to spread the weight of the wall as well as possible.

When you begin this process, it may feel like a mindfulness exercise in slowing down. Finding the right stone for the right part of your wall is like navigating the pieces of a three-dimensional jigsaw puzzle, where all of the puzzle pieces are different shades of grey. I can assure you that dedicating just a few hours to this art form will bring huge improvements to your level of skill and intuition. Stones, like all materials, are dispersing load. When you build a wall you are actually creating a load path in reverse.

Ingredients

– Stones of different sizes for stacking
– Small rocks and pebbles
– Loose fill – mud or cob mixture (see page 131)
– Optional – roll of string and wooden stakes

Method

1 Collect and lay out your stacking stones, ideally so that you can see most of them at once.

2 Clear the area on which you want to build your wall down to undisturbed earth. For this method, let us assume your wall is going to be a maximum of 3 metres high, i.e. the same height as the gable end of a house. In this case, the wall at its foundation will be at least 800mm wide. You will stack your stones so that your wall tapers inwards, evenly, narrowing at the top.

3 You might find it helpful to set up a horizontal string attached to two wooden stakes along the length of your wall. This can work as a reference to keep your stonework (somewhat) straight and level.

4 Start building your inner and outer walls, staying in line with your outer walls guide string. It is important to consider how each stone will disperse and transfer the loads placed on it. You are starting where the load will end and building upwards to where the future load will begin.

5 Ensure that each stone is resting on at least two other stones below it. This staggered pattern ensures maximum load dispersion and overall strength of your wall. Also ensure that every rock you place is stable and not wobbling. If it does wobble, you can add a small stone underneath it, known as a shim, to keep it stable.

6 You will need to incorporate through stones at regular intervals in the wall. A through stone connects the inner wall to the outer wall. We use them to tie the inner and outer walls together and to stop the two walls potentially falling away from each other. Depending on the size of your wall, the through stone could be placed at 1 metres, 2 metres and 3 metres high, at intervals of approximately 2 metres. You ideally want through stones at least halfway up, and the more you include, the more stable your wall will be.

7 Using your hands, fill in the cavity between your two walls with loose fill earth and smaller stones, known as hearting, or a cob mix.

TYRE POUNDING

This technique, developed and refined by Mike Reynolds and the Earthship Biotecture team, is still one of my favourite ways to build a structure. This free building material can be used as retaining walls in your garden, or the structural walls of your building. When used as a retaining wall, the wall's weight is not distributed vertically into the ground, but instead diagonally back into the hillside. For this reason, retaining walls often do not require a separate foundation; they are in themselves a foundation that can also serve as a wall.

From the outside, pounding tyres can look like unpleasantly hard work, but I can assure you that I have pounded tyres with people of all ages, with varying body types and strengths, and it is certainly achievable by anyone who puts their mind to it. Using the correct technique and understanding your own limitations is key to unlocking this economical and satisfying building technique.

Ingredients

– Disposed car tyres. Don't bother with truck tyres, bicycle tyres or wheelbarrow tyres. Car tyres are the most plentiful, and the easiest to work with
– Cardboard
– Sledgehammer. Depending on your arm strength, a sledgehammer between 3kg and 5kg is best. When I am pounding tyres all day, I like to use a 4.5kg sledgehammer
– Shovel
– Bucket
– Plenty of earth for filling tyres
– Spirit level
– (Ideally) a team of willing workers

Method

1 Clear an area of ground about 1 metre by 1 metre, removing all the topsoil until you reach undisturbed earth. Place your first tyre. Each fully pounded tyre weighs a hefty 130kg, so all tyres should be pounded in place – only minor adjustments can be made when it's full. Subsequent courses of tyres are staggered, like bricks or blocks, on top of the tyres below.

2 Line the bottom of your tyre with four layers of sturdy cardboard. This works as a temporary formwork holding the earth in place. It will no longer be necessary by the time it degrades.

3 Tyre pounding is easiest done in teams of two, one person filling the tyre with earth while the other compacts it with the sledgehammer. It should take a team of two people somewhere between 5 and 15 minutes to complete each tyre. Start by pushing or kicking the earth into the side walls of the tyre. Once the tyre starts to feel full, switch to your sledgehammer to start compressing the earth. The other team member can continue to add more earth as you compress it.

Note: The most effective way to pound a tyre is with many small compactions, rather than a few big compactions. You should never need to lift the sledgehammer above shoulder height. You are not swinging straight down into the ground but rather sideways, compacting the earth into the sidewall of the tyre.

4 Work your way evenly around the tyre. It might help to imagine you are working your way around a clock, starting at 1 o'clock, then 2 o'clock, then 3 o'clock, and so on, until you complete the circle. Ensure you are hitting only the dirt, and not wasting energy hitting the rubber of the tyre!

5 Once the tyre is starting to bulge at the top you can use the level to check it. It's important the tyre is level itself and level with its adjacent tyres. A fully pounded tyre will increase in height by about 40–50mm. Use your sledgehammer to compress any earth that is preventing the tyre being totally level.

6 Once you are satisfied that your tyre is level, finish by adding one more bucket of earth. Use the flat end of your sledgehammer to compress and level the last layer.

MAKING MORTAR

Mortar is the glue that holds our masonry together. If you plan to build masonry foundations you will need to understand its properties. It will take on various forms depending on the type of materials you are working with.

Essentially, mortar has three key ingredients: a binder, which holds everything together; an aggregate, which gives the mortar strength; and a liquid, water, which helps mix and integrate the ingredients, as well as being the catalyst for the curing/setting process. Common binders are cement, lime or clay. Aggregate is usually sand, gravel or fibre (natural fibres like straw/hair, or synthetic fibres). This process is much like making a cake batter, but on a much larger, and less delicious, scale.

When making mortar mixes we uses ratio to explain our measurements as opposed to fixed quantities. This allows the builder (you!) to scale the recipe as large or small as they need. With the same recipe you could make a cupful or a truckful of mortar.

Makes one wheelbarrowful of mortar

Ingredients
- 1 building bucket (approx. 10 litre capacity) of suitable binder. In this case we will use naturally hydraulic lime, which is available to buy in powdered form in bags, in various strength grades. An NHL 3.5 should work for most applications
- 3 building buckets of coarse sand
- 1 building bucket of water (approx.)
- Wheelbarrow
- A mixing tool. This could be a shovel if you are mixing a small amount by hand, or a cement mixer if you are mixing a large amount

Safety note: It is very important to wear protective equipment when working with lime, as it is a highly alkaline powder. Just like acids, highly alkaline substances are caustic and can cause serious irritation to the skin, eyes and lungs. Always wear thick rubber gloves, a suitable dust mask and eye protection when working with lime. If you feel any irritation on your skin when using lime, I find it very effective to rinse the affected area with vinegar and then to rinse with water. Vinegar is a highly acidic liquid which can counteract the lime's alkaline effects. But prevention is better than cure, and wearing gloves will protect you from lime burns.

Method

1 If you are making your mortar with a shovel or trowel, mix the lime and sand together until the lime is evenly dispersed throughout the sand. Once mixed, form the mound into a volcano shape – a cone with a well in the top.

2 Add some of your water into the well, gradually stirring in the dry ingredients from around it as it thickens. Try to keep the form of the volcano as you add more water and continue to mix the ingredients together until you have achieved a smooth and consistent mixture with no dry parts. There is no fixed ratio of water to dry ingredients here; you are looking for a porridge consistency, not so dry that it's crumbly, but not so wet that it runs off your shovel or trowel. You know it is the correct consistency if it easily holds its form.

3 Before you get to work using the mortar, clean all the tools you used to make your mixture, as cement and lime is much easier to remove when it is still wet! For the best and most consistent results, only make as much mortar as you can use within an hour. Mortar cures quickly, so it is better to make a series of small batches than a large batch which will require rehydration.

BRICKLAYING

Bricklaying or blocklaying is a straightforward yet effective skill when it comes to wall construction. Brick and block work are the natural evolution of stone work, and laying them comes with the efficiency and predictability of a repetitive task.

Brickwork is the star pupil of illustrating organised load dispersion. The same application process applies whether you are working with adobe bricks, red clay bricks or cement blocks, so learning how to bricklay will open up many possibilities for you in both your home and garden. The most important thing is to keep your brick courses/rows both level (straight from side to side) and plumb (straight up and down).

Bricks and blocks are heavy, and you might use hundreds of them in a wall. Ensure you are building on solid foundations that can carry and evenly disperse that kind of weight.

Ingredients

- Bricks (or blocks)
- Suitable mortar to match your bricks (see recipe on page 56)
- Wooden stake
- Club hammer
- Bolster chisel
- String
- Bricklaying trowel
- Small trowel or brick jointer tool
- Spirit level
- Hessian fabric (if required)

Safety note: Eye protection should be worn when cutting bricks. Gloves should be worn when working with lime mortar, to protect your skin from irritation.

Method

1 Mark out and draw a straight line onto your foundation, where your first course of bricks will sit. You can also use a length of timber for this. Using your bricklaying trowel, shovel a line of mortar along the marked line. Use enough mortar to make the line around 15mm in height, and slightly narrower than your bricks.

2 Place your first brick at one end of the line of mortar, and press down so that you end up with an even 10mm of mortar beneath the brick. Now apply mortar to one end of your second brick and place it in line with your first brick. Try to keep a consistent 10mm layer of mortar on all sides of the brick. Use your piece of timber or a spirit level as a straight edge, to ensure you are laying your bricks in a straight line. Repeat this step until you have reached the end of the wall and cannot fit another whole brick. Use your spirit level to check the height along the length of your wall, ensuring that your last brick matches the first, adjusting any bricks if necessary by pressing them gently into the mortar.

3 You will likely need to cut a brick to fit in at the end of your wall, or to start the beginning of the second course. To do this, first mark the desired cut line on your brick with a pencil or by scraping it with the corner of another brick. Place the brick on a firm surface, direct the chisel on to your scribe line and strike the head of the chisel with your hammer. If the brick does not break cleanly, flip the brick over and use the chisel and hammer to strike it again to remove any messy edges.

4 For the second course of bricks, use your trowel to lay a new bed of mortar along the top of the first three or four bricks from the first course. You want to create a slightly pointed, triangular shape with this mortar so it has a peak in the middle and room to compress at the sides. This is so that your next brick will have room to compress down to the right level when pressed down.

Use your spirit level and wooden layout stake to ensure you are at the right height and that your brick is level to itself, and square to the bricks below.

5 It can be useful at this point to set up a string line which will work as a guide to lay your bricks along. Place some mortar and lay a single brick at each end of the wall. Place a stick – or a mortar pin if you have one – into your mortar joint and tie your string to it. Run the string over the brick and along the length of the wall to the brick you placed at the end of your wall. Pull the string down over the brick, ensuring it is relatively taut, and tie it to the brick. This string should give you a level guide to where to lay your bricks across the wall.

6 You can now lay the remainder of the course using the string as a guide. Repeat the process until your wall reaches its desired height.

7 Once you have finished laying your bricks, run a small trowel or a brick jointer tool over your mortar joints to create a neat finish and fill in any gaps. If you are using a lime-based mortar, it is important to protect the mortar until it has cured/set. In hot weather, this typically involves covering the brickwork with a sheet of hessian fabric and regularly spraying the mortar with water so that it doesn't dry out too quickly and crack. In cold and wet conditions, hessian sheeting also protects the wall from excess water and extreme temperatures. It is not advisable to use lime mortar in freezing conditions, as it will not cure properly.

STEELWORK FOR CONCRETE

Great structures such as ancient colosseums were built from concrete without the use of steel, relying instead on the compressive qualities of concrete. However, the addition of steel into concrete provides us with the impressive tensile strength that many modern applications of concrete require. And while steel may be considered to be concrete's secret strength, it can also be its greatest weakness. Before diving into this recipe I suggest you read through 'The wonder and worry of concrete' on page 92 to familiarise yourself with how this material performs its tasks. Improperly placed steel can direct loads in the wrong direction, and can attract moisture, which will eventually cause it to rust and expand into the concrete, weakening it. So if you are going to add steel to your concrete, you must add it in the right way for it to be of benefit. The recipe given here is suitable for a single-storey timber frame structure.

Ingredients
– 10mm reinforcement bar (also known as rebar)
– Angle grinder
– Steel tie wire
– End cutters

Safety note: Always wear safety glasses to protect your eyes from sparks when using an angle grinder to cut steel. A grinder can be dangerous if misused, so please get someone to show you how to use this tool correctly if it is your first time. Ensure you wear suitable thick gloves when cutting steel, as it will heat up as you cut it.

Method

1 Draw up a cutting list for all of the lengths of steel you require. Your steel should always end 50mm away from the edge of the concrete surface, to reduce the chance of water reaching in and causing it to rust. For the same reason, the steel should not touch the ground, or the bottom of your formwork.

2 Use an angle grinder to carefully cut all the steel lengths you require. When using the angle grinder, remember that the tool's spinning motion is doing all the work, so you do not need to apply much pressure.

3 If you need to bend any of your rebar to turn corners or into pillars, you can use two steel tubes to create a lever, then bend by hand. Place the steel inside the tubes so that the part where you want the steel to bend is at the point where the two tubes meet. Pull the two poles towards you until the steel has bent to the angle you need. The longer your lever, the easier the steel will bend.

4 Tie your steel tightly together with steel tie wire using end cutters. Your tie needs to be as tight as possible without snapping the wire, so use the end cutters to twist and lever the steel tightly together. Practice makes perfect with this activity. All your steel should be around 100mm apart and in parallel lines, to create a suitable load path through the concrete.

5 You now need to suspend your steel within your formwork, ensuring it will remain in the correct position while the concrete is poured around it. This can be achieved with the use of tie wire or plastic risers, or small pieces of stone or concrete, to keep it off the ground.

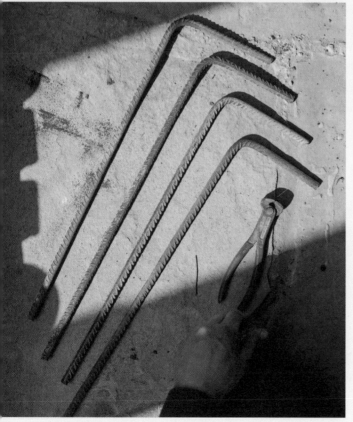

FORMWORK

We use what are known as forms to create shapes and contain liquid concrete until it dries. A form is essentially a vessel, or constraint, into which you pour your concrete. They can be made from a huge range of materials, as long as they are strong enough to hold the liquid concrete. Below ground level, foundation forms can be made of earth, and above ground they can be made from timber. Other suitable materials include plastic pipe or buckets. For concrete applications that require a smooth surface, such as a concrete countertop, you can use sheets of melamine to make your form. Melamine is a nitrogen-based, plastic-coated chipboard with a very smooth surface.

Remember, when it hardens, the concrete will show the surface of the formwork in every detail, including screws, scratches or dents. Such textures can even be a feature if you want them to be!

Makes a 1 metre long × 1 metre wide × 150mm deep form

Ingredients

– 150mm deep × 50mm thick pine, or similar softwood
– 4.5mm × 90mm screws
– Measuring tape
– Circular or hand saw
– Impact driver

Method

1 First, determine the dimensions that you want your finished concrete to be. These will be the internal dimensions of your formwork. In this example, we will create a concrete piece which is 1m × 1m × 150mm deep.
2 Cut two pieces of wood to form the internal sides of your concrete, 1 metre long each.
3 Cut two pieces of wood to form the other two sides of a square, long enough to cover the concrete and the wood you have already cut. In this example it will be the length of our concrete, which is 1000mm, plus the width of two pieces of 50mm thick wood, totalling 1100mm.
4 Screw the formwork together, remembering that you will need to be able to access these points easily to remove the screws once the concrete is dry. Place the form where you want your finished concrete to be. Take into account the level of the ground and any gaps in the formwork that may need to be filled.
5 Pour your concrete into the form.
6 After 24 hours, the concrete will be cured and you can carefully remove the formwork (or leave it, if this is your intention). The concrete is now what we call 'green', as it is still fresh and can be easily damaged or chipped. It will take up to 28 days for the concrete to be fully hardened but you can safely walk on and build on your concrete after five days.

REINFORCED CONCRETE

Again, I recommend reading 'The wonder and worry of concrete' on page 92 before attempting to make your own reinforced concrete. Understanding how steel and concrete relate to each other is essential for any problem-solving you may have to do in your own project. Reinforced concrete is a staple ingredient in conventional building, and while we develop affordable and sustainable alternatives it is a useful tool to have in your belt.

When we are working with concrete, we talk in ratios, or parts of a whole. This is so that we can make a bucket or a truckload of concrete with the same recipe. Depending on the strength of concrete you are trying to achieve, you might add more binder (cement) or aggregate (gravel and sand), but this is a safe, standard recipe for the majority of applications.

Safety note: Concrete is a very fine powder which can become airborne easily. It can cause irritation to the eyes and the lungs, so always wear glasses and a dust mask when handling the powder. Always wear gloves when working with concrete, as it will dry and irritate the skin. It can cause burns when in prolonged contact with the skin, since it is derived from lime, which is almost pure alkaline. If you do suffer any mild irritation of the skin due to contact with the lime, it can help to cover the area with vinegar, which counteracts the alkaline burn with an acid, then rinse well with water.

Ingredients

- 4 parts gravel aggregates
- 1 part cement binder
- 2 parts sand aggregates
- Clean water – water is the one part of the concrete mix that you will never see a ratio for, but is essential in the process. The amount of water needed for your mix depends on a number of factors, and you will need to trust your intuition, or get some advice from a more knowledgeable friend, until you get the hang of it. The amount of water you need will depend on the humidity, or whether it rained the night before, so the best approach is to aim for a suitable consistency as opposed to applying specific quantities. You are aiming for a runny porridge consistency, where the water is fully mixed in with the rest of the ingredients.
- Wheelbarrow
- Spade
- Optional – cement mixer

Method

Hand mixing method

1 Small amounts of concrete can easily be mixed by hand, in a wheelbarrow, using a spade. It is best to mix the dry ingredients together first, and then slowly add the water until you reach the consistency you are looking for. Keep an eye on the consistency of your mix throughout the entire process, adding water, little by little, as you need it. Go easy – a little water can make a big difference.

Cement mixer method

1 Add a little water to the mixer first, followed by all of the gravel aggregate to create a crunchy mix. This ensures that your cement is dispersed evenly throughout the mix when added, instead of clumping together.

2 Now carefully add the cement powder to the mixer, standing back to avoid the dust. Let it mix for about a minute or until fully combined.

3 Finally, add the sand. Allow the mixer to spin for at least 5 minutes to ensure all the ingredients are properly dispersed and wet. Allow your mix to turn for another 5 minutes while you refill your buckets of sand and gravel and get your wheelbarrow ready to pour the mix into.

Concrete is much easier to clean and remove while still wet. As soon as you have finished using a tool that has concrete on it, wash it thoroughly. If you do suffer any mild irritation of the skin due to contact with the cement, I find it effective to rinse the area with vinegar, and then rinse with water. The acidic vinegar helps to counteract the cement's high alkalinity.

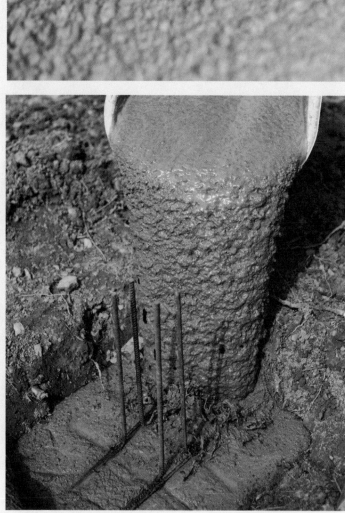

CHAPTER 2

STRONG MATERIALS: GARBAGE AS BUILDING BLOCKS

left Recycled bottle wall in an Argentinian primary school.

below Potential building materials.

opposite Turning trash into treasure.

I had a life-altering realisation a number of years ago when I was working on the construction of an off-grid primary school in a small coastal fishing village in Uruguay. Walking home from the Escuela Sustentable job site each day, I noticed that the majority of the houses I passed had been constructed using an array of materials, such as timber, steel and brick, combined with natural approaches, such as cob and straw bale. Even more interesting was the fact that nearly all of the structures were partially made of repurposed garbage. I learned that the building code in that particular municipality required no permits for domestic dwellings and so people had a degree of freedom to build their own unique homes.

Just as garden birds create their nests from whatever they find each spring, these owner-builders had built with whatever usable materials they could get their hands on. Shipping containers were adapted to become living spaces, with earthen walls protruding from them on foundations of repurposed concrete pavements. Neat piles of what could be considered garbage were separated and carefully organised according to their material, size and colour.

These innovative homes were undoubtedly a personification of their makers, who, unbridled by restrictive planning regulations or huge loans, had creatively built within their means. There was a lovely freedom to their approach, and the houses I observed were surprisingly well built. Clearly knowledge and skill had been shared between them, so that each generation of homeowners learned

from the last. I am sure the houses had their issues, perhaps a leaking window, or a door not closing quite right; there was likely room for improvement in efficiency and function. But these homes had the advantage of being built by their inhabitants. This meant that the challenges and costs of any home improvements were in the hands of the owners, the builders and the inhabitants – all of which were the same people.

Observing these Uruguayan homes led me to question why we don't build all our structures with the free materials we have access to. The answer to that is multi-faceted, but it starts with education, inspiration and the motivation to gain a deeper, more coherent understanding of how materials work.

An unfortunate truth

It is difficult to overstate the harrowing effects that the construction industry is having on our environment. This global industry consumes one-third of the world's mined resources, while at the same time producing a quarter of the world's solid waste. Data from a 2021 geological survey by the US Minerals Education Commission shows that in their lifetime, the average US citizen requires 17,462kg of raw minerals to be mined to maintain their current standard of living. The top four minerals on the list are stone (4275kg), sand and gravel (3069kg), lime (297kg) and clay (65kg). These four minerals are the primary ingredients of concrete, our most beloved and destructive building material. In his fascinating book *The World in a Grain*, Vince Beiser reflects on the concrete revolution and our massive

opposite, top Straw as a building material.

opposite, bottom Salvaged Australian jarrah.

far left Typical coarse sand for concrete.

left Typical gravel for concrete.

dependency on this material. Almost non-existent in our buildings 120 years ago, it is hard to believe that we as a people now annually consume the same amount of concrete it would take to build the equivalent of eight New York Citys.

A culturally inherited desire for shiny new buildings, coupled with stringent insurance obligations, too often encourage us to demolish and rebuild, instead of improving and protecting our older buildings. I believe that, with humankind's incredible creative capacity, we could find a multitude of ways to save the energy of today by building on the efforts of yesterday. It seems beyond logic that we can justify knocking down 200-year-old structures that have, quite literally, stood the test of time. This linear approach to construction, where our resources are on a one-way journey from cradle to grave, is having an irrevocably destructive impact on our planet.

The expectation of having everything we need or desire available to us at the click of a button has added pressure to our already strained systems. More consumption means more stock, which means more storage, more buildings, more electricity, more fuel and more mining of precious, irreplaceable resources, all leading to us creating more waste! It is easy to pass the blame on to some other point of the cycle, but I believe we must consider the aspects we have control of before demonising those beyond our control. Consumer demand drives production demand, which in turn drives the creation of new materials. Similarly, consumer demand for the lowest price possible drives the production industries to create things as cheaply as possible, which ultimately leads to greater amounts of waste. The idea that it is most often cheaper to replace any item than to fix it is a worrying notion.

It should come as no surprise that profits guide our industries. In this competitive, capitalist society it is difficult to create jobs and build infrastructure around concepts that are, in their essence, about spending less money. Repurposing waste material saves money, which is why it is a good idea, and this is also why there is little to no industry currently around it. If there were, it's unlikely that these materials would still be affordable! The certification process that new products must go through can be very lengthy and expensive, often requiring hundreds of hours of rigorous independent testing. The hours and money required are easily justified if the product ends up going to market and offsets the initial investment. There are very few people who are willing and able to invest the time and energy required to certify waste materials for the construction process, given that they will never make any money back off this process. But they are out there. Mike Reynolds, for example, has published an extensive report ('Seismic Performance Evaluation and Tire Construction Analysis', 2017) on the performance evaluation of repurposed tyre construction, and has laid the path for others to follow in bringing repurposed materials to the forefront of the construction conversation.

Is recycling the solution?

Before throwing all our recyclable waste into the nearest green bin, we need to consider what the recycling process is really costing the planet. The recycling industry has come under scrutiny in the past decade, as researchers have uncovered holes in the system. How far is our rubbish having to travel and how much energy is being used to transform it back into a useful form? The justifications of single-use products are quickly evaporating in many aspects of our lives, and it can be hard to understand why we are still producing materials which cannot be easily recycled into a useful form for reuse.

To understand how the recycling solution is not always the best solution, let's follow the post-use journey of a milk carton. Made from Tetra Pak, a laminated mix of cardboard and film, this is actually a rather difficult material to recycle. There is only one facility in all of the UK and Ireland which recycles Tetra Pak; it's located in Fife in Scotland. Fife is 750km from my home, and as I look at a nut milk carton on my kitchen table, I try to work out how I can justify the transport required to package and recycle an item that is only in my house for a matter of days. So you can see how, while the carton is technically recyclable, the fuel and energy required to do so hardly justifies its reuse.

More widely recycled materials such as glass, steel and aluminium are required to go through an intense smelting process, much like the process first used to create them, to get them back on our shelves. Certainly, it is a positive that recycling these

Shadows playing over piles of sand in Australia.

materials reduces the amount of new minerals being mined. However, much like the milk carton, the energy required to recycle them is equal to the energy required when they were first created. As our fossil fuel reserves around the world are depleted we may be forced to question whether recycling everything is really a feasible solution for our consumption problems.

A consumer-led demand for more transparency across supply chains has sparked the introduction of the circular economy model. Cradle-to-cradle certification systems have certainly had a positive impact, creating incentives for product developers and manufacturers of construction materials. However, we need more than a step in the right direction to change the trajectory of this global industry. We need to acknowledge it as a crisis, and act accordingly. The ceaseless demand for new buildings, and the expansion of cities and transport routes makes it seem improbable that we could interrupt the flow of the new construction river.

I have a theory that this is largely due to a case of inertia. Inertia is defined in the Oxford English Dictionary as 'a property of matter by which it continues in its existing state of rest or uniform motion in a straight line, unless that state is changed by an external force'. In the construction industry, we are on a trajectory that has been gaining momentum for decades. The more momentum and speed an idea has, the greater the force must be to break our patterns, and challenge the status quo.

Imagine the scale of the external force required to disrupt something as momentous as our global construction industry. The mining, the furnaces and the trucks, all moving in endless loops to supply our never-ending consumer demand, would require something huge to change. But change is possible, when our species is under threat. The coronavirus pandemic can be seen as the external force that almost completely stopped the worldwide travel industry nearly overnight. We will always remember the period of completely clear blue skies when our diesel-guzzling planes were grounded due to the virus.

I feel excited when I hear of the incredible innovators out there, turning industry waste products into construction materials. There are creators growing furniture out of fungi and sawdust, and making wall panels out of end-of-life paper cellulose. Someone has even found a use for the invasive and destructive Japanese knotweed plant, combining the fly ash from incinerating it with crushed crayfish shells to create a concrete substitute. We read all about these incredible innovations while we wait for the concrete truck to arrive to pour our new driveway, and wonder at what the future might look like, filled with all of these amazing ideas. The only thing standing between our present and the future is the inertia of our current destructive trajectory. Perhaps we are just like all other forces of nature in that we choose the path of least resistance. If we can create systems that encourage us to choose greener, cleaner, more conscious consumption and construction habits – because they are actually the easiest and most efficient (or affordable) option – then, I believe, real change is possible.

A need for solutions

The ultimate solution to the crisis primarily lies in reducing our demand for new buildings, and therefore creating less materials to start with. From the home designer's perspective, we should be facilitating a discussion about smaller spaces, more humble structures and challenging the norm of daily living practices and expectations. So many of the houses dotted across the landscape are based on a design from a time before our pressing environmental and energy crisis. These unnecessarily large buildings are often highly energy consumptive and built from materials whose embedded carbon footprint is just not considered. Perhaps it is time to redesign the way we live and

interact with our buildings in a way that reflects the current climate and needs of the environment.

The energy required in the recycling process could be heavily subsidised by a stronger focus and infrastructure around repurposing waste materials. It would require minimal additional energy to create a resource out of what was once seen as a burden. If we could focus our energies on turning the solid waste created by our global industries back into a usable construction material, we would drastically reduce our reliance on mined material, whilst significantly reducing the amount of waste that is being created.

The way we can assist in bringing this somewhat controversial idea to life is through education, independent thinking and creativity in our day-to-day lives. When exploring ways to repurpose materials I have found it most effective to look outside my industry. It is not such a stretch to imagine repurposing construction waste into a construction material. It gets a little more interesting when we consider the waste from other aspects of our lives. Think of the materials that you bring into your home every week and the materials you are sending out in your bins for recycling or to landfill. What are they made of?

left Salvaged wood for reselling.

Exercise
BEAUTY IS IN THE EYE OF THE BEHOLDER

bottom left Masonry wall with aluminium cans as building blocks.

opposite Masonry wall with wine bottles as building blocks.

Let's explore the qualities of some of our most commonly used materials. Go to your recycling bin and pull out a few things to observe.

ALUMINIUM CANS

Qualities: The cylindrical aluminium can is designed to hold a pressurised liquid, to be waterproof and to resist punctures or damage during transportation.

Recycling potential: Aluminium is comparatively easy to recycle, in that it has a melting point of 660°C. Around 70% of aluminium cans sold in Ireland make it to the recycling stage and are shipped overseas to be melted down. The smelting process for new and recycled aluminium produces highly combustible off-gases that must be carefully handled.

GLASS BOTTLES

Qualities: Allows for strong shapes which ensure that this fragile material can survive transportation. A highly waterproof material, available in an assortment of colours.

Recycling potential: Glass is widely recyclable, with a melting point of 1400–1600°C. After being colour-sorted it can be crushed and transported to a melting plant. Clear glass is the most valuable, while brown glass is the least valuable. Much glass is simply crushed back into a sand-like material for other uses.

CORRUGATED CARDBOARD

Qualities: A flexible, easy-to-cut material which is insulative. Corrugated cardboard has an R-value of 3 per inch, compared to fibreglass batt insulation, which has an R-value of 5 per inch, making cardboard around 60% as effective.

Recycling potential: Cardboard recycling requires shredding and pulping to break down the material into fibres, which can be recycled into other paper products.

In pointing out the useful qualities of these materials in their 'waste' form, I hope to illustrate their value and show that finding ways to repurpose materials will have a far greater impact in protecting our environment than sending everything through the recycling process.

Changing the way we see trash

In 2017, friends and I built a sauna in an old stone ruin on my property. We repurposed damaged building materials and insulation from the main cottage to create the majority of the structure. We also collected over a thousand coffee bags from cafés around Dublin. These bags were made of a laminated foil and plastic material which is extremely hard to recycle. We ironed them together to make shingles and lined our sauna with them as a reflective barrier for the heat in the sauna and as a moisture barrier before the final cedar was installed.

Seeing trash as a potential building material might seem controversial from a 21st-century perspective, but if you were to arrive on Earth today and took a look around, you wouldn't be blamed for assuming that trash is a naturally occurring material. We have piles of waste we don't yet know what to do with, mountains of used car tyres visible from space as soft plastics blow across our remote landscapes. And too often, the best solution we can come up with is to either bury or burn it, so that we can make room for more trash! If we can step out of the traditional construction mindset for a moment and consider the inherent value of the waste materials we are surrounded by, we may still have a chance to redirect this flow to a better outcome. Of course, many of our current 'building with trash' approaches are in their infancy, and could be evolved to become more efficient and easier to implement. However, it must be acknowledged that many of the materials that travel through our

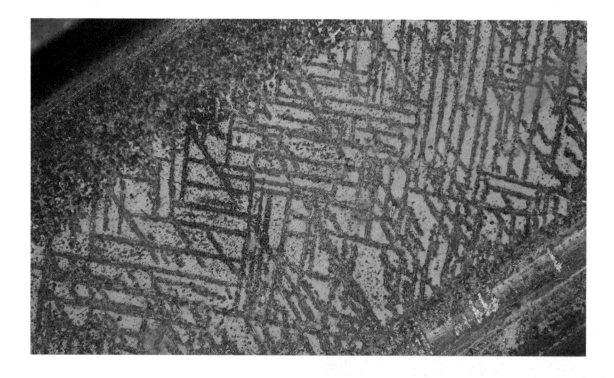

What makes a material strong?

household bins are engineered and designed in such a way that they can handle much greater stresses than we will ever place on them in a structure. I would love to see what would happen if we were to put even a fraction of the creativity and ingenuity at play when we are searching for new materials into finding useful and practical ways to repurpose what we already have.

What image comes to mind when you are asked to imagine a strong material? Do you envisage something rigid and immovable, such as concrete or the bricks your home is made of? Or perhaps you see a length of steel stretching far across a void? No matter what material comes to mind, understanding the characteristics of strength enables us to explore the potential use of objects that might otherwise be considered to be waste.

Materials should not be compared with each other by their strength alone, but rather by what makes them strong. To define strength we must uncouple it from its components. The strength of a

opposite Celebrating the ageing process of steel.

right Clay bricks with cement mortar on a stone foundation.

far right Exposed, perfectly laid cement blocks with cement mortar.

material is the potential stresses it can carry, combined with the strain it might experience under these stresses.

Stress can be understood as the amount of force being applied to a material over a certain area, whether that force is gravity, a car or an elephant. Strain can be understood as how much a material stretches or compresses under a given stress.

If we divide the stress by the strain we can understand a material's rigidity or elasticity. The strength of the material can be determined by how much force is required to break it.

Basically, all materials fall somewhere on a scale from rigid to elastic. A brick carrying a 20kg load might be considered rigid, while a fishing line suspending the same load is elastic, despite them both supporting the same weight and thus having a similar relative strength. It's good to remember that all structures are made of a considered combination of materials. Many parts making a whole. In the same way as all materials are made up of a combination of atoms, many parts making a whole, the structures we build on a human scale are versions of the structures we find naturally all around us.

right Discussing cladding options for a rounded, sloping structure.

opposite Elegant, load-bearing timber structure.

Understanding span

The span is the gap between two supporting elements. Understanding span is critical to managing load displacement in our buildings. It allows us to determine the grade, dimension and quantity of timber necessary to safely achieve our supporting structure. A span chart will help us to work out how many roof rafters or ceiling or floor joists we will need to cover a certain area, as well as the specific thickness and height these planks must be.

There are numerous free span charts available online, and understanding the language of these charts makes them a useful reference. It is important to ensure that you are referring to a span chart that is relevant to your location, material and end use.

Loading	Permanent (gk) 0.30 kN/m						Imposed UDL (qk) 1.50 kN/m			Imposed Concentrated (Qk) 2.00 kN					
	Strength class of timber														
Target size (mm) (bxh)	C14			C16			C18			C24			C27		
	Spacing of joists (m)														
	300	350	400	300	350	400	300	350	400	300	350	400	300	350	400
	Maximum clear span between supports (m) – 50mm allowed for bearing														
38 × 100	1.56	1.55	1.51	1.73	1.71	1.70	1.84	1.82	1.80	2.03	2.00	1.98	2.07	2.05	2.02
38 × 115	1.98	1.96	1.94	2.12	2.09	2.07	2.24	2.21	2.19	2.47	2.44	2.41	2.52	2.49	2.45
38 × 125	2.23	2.20	2.18	2.38	2.35	2.32	2.52	2.49	2.46	2.77	2.73	2.68	2.82	2.78	2.72
38 × 150	2.88	2.84	2.76	3.06	3.02	2.89	3.23	3.14	3.01	3.54	3.36	3.22	3.59	3.41	3.26
38 × 175	3.54	3.36	3.22	3.70	3.52	3.37	3.85	3.67	3.51	4.04	3.90	3.75	4.08	3.94	3.81
38 × 200	3.99	3.84	3.68	4.12	3.98	3.85	4.24	4.10	3.98	4.45	4.30	4.17	4.49	4.34	4.21
38 × 225	4.35	4.20	4.08	4.49	4.34	4.21	4.63	4.47	4.33	4.85	4.69	4.55	4.89	4.73	4.59
44 × 100	1.74	1.72	1.71	1.86	1.84	1.82	1.97	1.95	1.93	2.17	2.15	2.12	2.22	2.19	2.17
44 × 115	2.13	2.10	2.08	2.27	2.24	2.22	2.40	2.37	2.34	2.64	2.60	2.57	2.69	2.66	2.62
44 × 125	2.39	2.36	2.33	2.55	2.52	2.48	2.70	2.66	2.63	2.96	2.92	2.81	3.01	2.97	2.85
44 × 150	3.07	3.03	2.90	3.27	3.17	3.03	3.45	3.30	3.16	3.71	3.53	3.38	3.76	3.58	3.43
44 × 175	3.71	3.53	3.38	3.86	3.69	3.54	3.98	3.84	3.68	4.17	4.03	3.91	4.21	4.07	3.95
44 × 200	4.13	3.99	3.86	4.26	4.12	3.99	4.39	4.24	4.11	4.60	4.45	4.31	4.64	4.49	4.35
44 × 225	4.50	3.34	4.22	4.64	4.49	4.35	4.78	4.62	4.48	5.01	4.84	4.70	5.05	4.88	4.74
47 × 100	1.80	1.78	1.76	1.92	1.90	1.88	2.04	2.01	1.99	2.24	2.21	2.10	2.29	2.26	2.23
47 × 115	2.20	2.17	2.14	2.34	2.31	2.28	2.48	2.45	2.42	2.72	2.68	2.65	2.77	2.74	2.68
47 × 150	3.17	3.09	2.96	3.36	3.24	3.10	3.54	3.37	3.23	3.79	3.60	3.45	3.82	3.65	3.50
47 × 175	3.79	3.61	3.45	3.92	3.77	3.61	4.04	3.90	3.76	4.24	4.09	3.97	4.27	4.13	4.01
47 × 200	4.19	4.05	3.92	4.32	4.18	4.05	4.45	4.30	4.17	4.66	4.51	4.38	4.70	4.55	4.42
47 × 225	4.56	4.41	4.28	4.71	4.55	4.42	4.84	4.68	4.55	5.08	4.91	4.77	5.12	4.95	4.81
75 × 175	4.21	4.07	3.95	4.34	4.20	4.08	4.47	4.32	4.20	4.68	4.53	4.40	4.71	4.57	4.44
75 × 200	4.63	4.48	4.35	4.77	4.62	4.49	4.91	4.76	4.63	5.14	4.98	4.85	5.18	5.02	4.89
75 × 225	5.03	4.87	4.74	5.19	5.03	4.89	5.34	5.17	5.03	5.58	5.41	5.27	5.62	5.46	5.31

Source: https://www.housebuild.ie/construction/building-guidelines/roof-level/roof-timber

Loading refers to the quantity and type of load that the joists are designed to carry. You can see that these have been broken up into three main categories in the chart, with each category associated with a specific weight per metre. As discussed in Chapter 1, permanent or dead loads are the loads that do not change in the building. They often refer to the weight of the structural materials such as roof tiles, floorboards or internal walls. Imposed or live loads are the loads that are added to a building, such as people, appliances and furniture.

Strength class of timber refers to the suggested grade of timber permitted for use. Before a tree is milled, a certifier will examine the end grain of the timber and assess the species and grade. They then divide it into different strength classes. The higher the grading, the stronger the timber and the greater the distance it can span.

Spacing of joists refers to the distance between your joist members. This is measured between the centres of each joist, rather than the space between them. The more often your joists occur within a space, the greater the number of joists and thus the more elements there are to disperse the loads. **Target size** refers to the dimensions of the joists. Joists are always placed in a vertical orientation, so they should always be taller than they are wide. This chart, and most charts like it, will show a variety of suitable timber dimensions. This allows you to work with the timber you have available, or in a way that suits your project more specifically, for example leaving room for plumbing and electrical services to fit into a floor cavity.

Maximum clear span is the maximum distance the joist can carry the load.

Once you understand what all these elements are referring to you should be able to determine what your options are when looking to span a certain distance. This table can be used in both directions, so if you already have the timber you plan to use, the table will show you how far apart your joists will need to be. Similarly, if you have already decided on the span, the table can show you which dimension of timber will be suitable. You can see this method in action in the exercise below:

above Exposed structural elements in Connemara, Ireland.

Exercise
WHICH WOOD COULD?

Use the table on page 88 to determine the most suitable options to span a distance of 4.7m (4700mm). Anything graded to span 4.7m or greater will work, so we must first look through the span options and find all the numbers bigger than 4.7m. Once we have narrowed down these options, we can look across the chart to see which one use the cheapest option of wood. Which will likely be our second highest priority, after safety. Both the size of the timber and the quantity will affect our costs, and it's useful to note that it may, at times, work out cheaper to use a smaller quantity of a larger dimension of timber, rather than a greater quantity of a smaller dimension.

Assuming you have not yet purchased your lumber, it's best to approach this table somewhat in reverse. You know that you need to span a distance of 4.7m, so scan through the numbers in the Maximum clear span portion of the table until you find 4.7m or slightly higher.

You'll see that several options are possible:

Option 1: Using wood with the dimensions of 47 × 200 and graded at C27, you will have to space them at 300mm intervals.

Option 2: Using wood with dimensions of 47 × 225 and graded at C16, you will have to space them at 300mm intervals.

Option 3: Using wood with dimensions of 75 × 225 and graded at C14, you will have to space them at 400mm intervals.

So now that you know your options when it comes to safety, you must determine what the cheapest way to span this distance will be. There are a few factors to consider here. The larger a timber's dimensions, the more it costs. Also, the higher a timber's grading, the more it will cost. So it's important to note that the increments you will need to place them at determine the quantity you will need to buy, and will have a significant impact on the overall cost as well. The best way forward from this point is to call your lumber supplier and ask for a quote on these three items and see which will work out the cheapest for you to safely build your structure.

Loading	Permanent (gk) 0.30 kN/m						Imposed UDL (qk) 1.50 kN/m						Imposed Concentrated (Qk) 2.00 kN		
	Strength class of timber														
	C14			C16			C18			C24			C27		
Target size (mm) (bxh)	Spacing of joists (m)														
	300	350	400	300	350	400	300	350	400	300	350	400	300	350	400
Maximum clear span between supports (m) – 50mm allowed for bearing															
38 × 100	1.56	1.55	1.51	1.73	1.71	1.70	1.84	1.82	1.80	2.03	2.00	1.98	2.07	2.05	2.02
38 × 115	1.98	1.96	1.94	2.12	2.09	2.07	2.24	2.21	2.19	2.47	2.44	2.41	2.52	2.49	2.45
38 × 125	2.23	2.20	2.18	2.38	2.35	2.32	2.52	2.49	2.46	2.77	2.73	2.68	2.82	2.78	2.72
38 × 150	2.88	2.84	2.76	3.06	3.02	2.89	3.23	3.14	3.01	3.54	3.36	3.22	3.59	3.41	3.26
38 × 175	3.54	3.36	3.22	3.70	3.52	3.37	3.85	3.67	3.51	4.04	3.90	3.75	4.08	3.94	3.81
38 × 200	3.99	3.84	3.68	4.12	3.98	3.85	4.24	4.10	3.98	4.45	4.30	4.17	4.49	4.34	4.21
38 × 225	4.35	4.20	4.08	4.49	4.34	4.21	4.63	4.47	4.33	4.85	4.69	4.55	4.89	4.73	4.59
44 × 100	1.74	1.72	1.71	1.86	1.84	1.82	1.97	1.95	1.93	2.17	2.15	2.12	2.22	2.19	2.17
44 × 115	2.13	2.10	2.08	2.27	2.24	2.22	2.40	2.37	2.34	2.64	2.60	2.57	2.69	2.66	2.62
44 × 125	2.39	2.36	2.33	2.55	2.52	2.48	2.70	2.66	2.63	2.96	2.92	2.81	3.01	2.97	2.85
44 × 150	3.07	3.03	2.90	3.27	3.17	3.03	3.45	3.30	3.16	3.71	3.53	3.38	3.76	3.58	3.43
44 × 175	3.71	3.53	3.38	3.86	3.69	3.54	3.98	3.84	3.68	4.17	4.03	3.91	4.21	4.07	3.95
44 × 200	4.13	3.99	3.86	4.26	4.12	3.99	4.39	4.24	4.11	4.60	4.45	4.31	4.64	4.49	4.35
44 × 225	4.50	3.34	4.22	4.64	4.49	4.35	4.78	4.62	4.48	5.01	4.84	4.70	5.05	4.88	4.74
47 × 100	1.80	1.78	1.76	1.92	1.90	1.88	2.04	2.01	1.99	2.24	2.21	2.10	2.29	2.26	2.23
47 × 115	2.20	2.17	2.14	2.34	2.31	2.28	2.48	2.45	2.42	2.72	2.68	2.65	2.77	2.74	2.68
47 × 150	3.17	3.09	2.96	3.36	3.24	3.10	3.54	3.37	3.23	3.79	3.60	3.45	3.82	3.65	3.50
47 × 175	3.79	3.61	3.45	3.92	3.77	3.61	4.04	3.90	3.76	4.24	4.09	3.97	4.27	4.13	4.01
47 × 200	4.19	4.05	3.92	4.32	4.18	4.05	4.45	4.30	4.17	4.66	4.51	4.38	4.70	4.55	4.42
47 × 225	4.56	4.41	4.28	4.71	4.55	4.42	4.84	4.68	4.55	5.08	4.91	4.77	5.12	4.95	4.81
75 × 175	4.21	4.07	3.95	4.34	4.20	4.08	4.47	4.32	4.20	4.68	4.53	4.40	4.71	4.57	4.44
75 × 200	4.63	4.48	4.35	4.77	4.62	4.49	4.91	4.76	4.63	5.14	4.98	4.85	5.18	5.02	4.89
75 × 225	5.03	4.87	4.74	5.19	5.03	4.89	5.34	5.17	5.03	5.58	5.41	5.27	5.62	5.46	5.31

The wonder and worry of concrete

We have already explored the impressive load-bearing capabilities of concrete, but it seems pertinent to explore the world's most popular and controversial building material a little deeper. Before we begin, let us first clarify the relationship between concrete and cement.

below Exposed concrete brickwork.

opposite Structurally reinforced concrete subfloor and solid concrete wall with steel left exposed for further concrete work.

Cement is an extremely fine, grey powder, with which we make all cement-based products. It is derived primarily from limestone, combined with a range of other minerals.

Cement mortar is a mixture of cement and sand, which is typically used for joining brick- or blockwork, and setting paving slabs. A typical ratio for the mortar is 1:5 cement:sand.

Cement plaster is a mixture of cement and sand, with more water added to it than mortar, so that it can be spread thinly with a trowel onto vertical surfaces. A typical ratio for cement plaster is 1:4 cement:sand.

Concrete is made up of three ingredients: cement, aggregate (in the form of sand and heavier gravel) and water. Concrete is typically used for foundations. A typical ratio for concrete is 1:2:4 cement:sand:gravel.

Water serves to activate the binding qualities of cement and turn the powder and aggregate into a liquid. This liquid can then be poured, plastered or trowelled into any form or shape, depending on its viscosity, and will set in that form.

There are many additives that can be added to concrete, including reinforcing fibres, which add a structural matrix to the mix, or accelerants and liquid latex, which speed up the drying process. Such accelerants are useful in cold climates, where concrete must be prevented from freezing before it has fully cured/set. Most often, when we talk about concrete these days, we are usually referring to reinforced concrete, which has one more added ingredient: steel reinforcement bar. This textured

metal rod is embedded into the concrete while it is still in liquid form, to give it additional strength. You may recall from Chapter 1 the description of concrete as a man-made stone. The combination of different-sized aggregates held together with the cement binder gives it an incredibly high compressive strength, meaning that it can handle a large load with minimal deflection. So what changes in the sand and gravel to make it that much stronger in its concrete form? Again, the principle of organised load distribution applies. Unlike pure water, the density of the cement mixture (referred to as 'mud' in the trade) prevents the sand and gravel settling to the bottom, as it would normally do in a liquid. The evenly distributed, varying sizes of aggregate are suspended in the cement slurry. Once it has cured, or dried, we are left with our stone-like material. The aggregate acts like pebbles in a stream – the load pushes down through the concrete, and every time it encounters a piece of aggregate, it is divided and distributed wider and wider throughout the block. Loads are divided among the largest stone aggregate, then again by the smaller sand aggregate, and this process repeats itself, spreading our point load over a wider footprint. You will see in this illustration that the uppermost corners of the block are not receiving any of the load. This is the most efficient way to design footings, following the load path and using only as much concrete as you need.

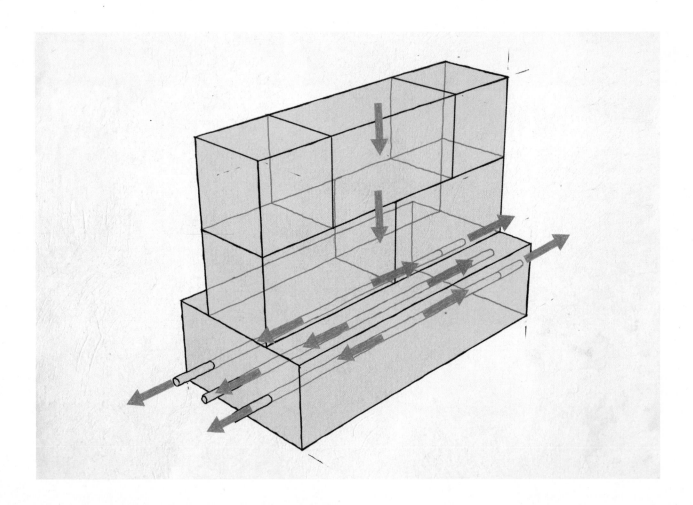

Concrete's impressive compressive strength has allowed us to repetitively manufacture regular-sized blocks with a reliable load capacity and a long lifespan. Concrete blocks, like red clay bricks, can be used in many ways around our structures, but they do have their limitations. Concrete without steel is extremely weak in its tensile strength, which means that it has very little elasticity. Its ability to absorb a load, disperse that load and bounce back to its original shape is limited. When we use a material to span a gap in a flat, level way and then place a load on it we are asking it to be both rigid and elastic at the same time. Unlike an arch, which is only experiencing compression under a load, this flat beam is experiencing both. We understand through Hooke's law that every material has a reaction to a load, even if that reaction is imperceptible to the human eye.

When a material deflects under a load, the top of the material experiences compression, while the bottom of the material experiences tension. This is why it is not only important to understand where the load will travel, but whether your chosen material is likely to allow it to do so.

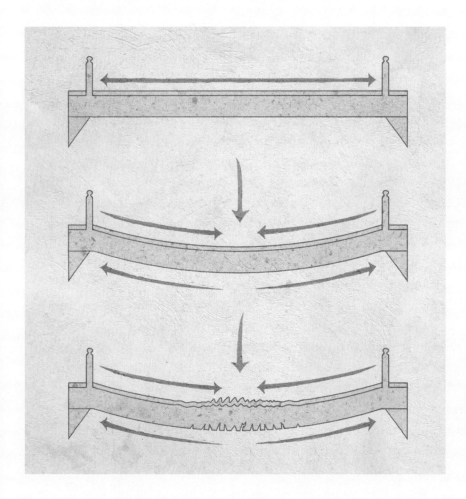

opposite Strip footing with arrows illustrating the load path.

left The effect of load on a given material (in this case, concrete), highlighting the potential duality of compression and tension.

By adding lengths of steel into our concrete, we offer the load something flexible to travel along. By doing so, we successfully combine a compressive material with a tensile material. The steel is twisted or notched, so that it bonds well to the concrete and provides the necessary elasticity and tensile strength that the concrete would otherwise lack. We can now span great distances with concrete and pour much thinner slabs. Any point load placed on a slab now only needs to find the steel reinforcing bar to distribute its weight. Because the steel is added for its tensile strength, it is important to use it in the part of the concrete which experiences tensile strain.

Steel and concrete have almost identical expansion and contraction rates under temperature changes. This means that the concrete does not crack when increased temperatures cause the steel within it to expand. Instead, the concrete expands along with it, at the very same speed.

You will have to forgive my excited tone as I talk about the ingenuity of concrete. Hopefully by now you can see how its invention was an amazing amalgamation of the concepts of rigidity and elasticity. It is a material that brings some of our most far-fetched dreams to reality. So revolutionary was the creation of concrete that it has become one of our most depended-upon resources, and sadly, like so many resources we overuse, its production is having a devastating effect on our planet.

Concrete's rise to fame was largely due to its core ingredients being easy to mine and seemingly in plentiful supply. In his book *The World in a Grain*, Vince Beiser reports on the destructive and illegal process of mass sand harvesting. Sand mafias and sand-related murders are connected to the ever-increasing demand and dwindling supply of sand. Very little of the world's natural sand is suitable for concrete, which requires sand granules with sharp edges, to ensure they lock together well in our concrete mix. The city of Dubai, which borders one of the world's largest deserts, imports its sand from as far away as Australia to build its epic skyscrapers.

In the year 2020, the energy that was consumed in crushing and furnacing minerals to make cement produced more than 4 billion tonnes of CO_2 emissions. Cement is the primary recipient of the tens of billions of tonnes of sand that is being mined annually. We are building cities faster than ever before, and we are only just starting to consider the true costs of this. Indeed, our castles are still very much made from sand.

Despite understanding the negative environmental impact of cement production, our worldwide demand for it continues to rise. We choose it because we know how to use it, and it's as if we have forgotten how we survived without it. Cement was invented a mere 200 years ago, and in just a few generations, after centuries of construction, we have forgotten the path we took to get here. I would wager with great confidence that if cement production stopped today, we would have our solution tomorrow, and the world would be a better place for it. Humans are incredibly adaptable creatures and I believe we would see a tremendous surge of innovation in the construction industry, given the right motivations. If the global coronavirus pandemic taught us anything, it is that when we are under threat, we are instinctively reactive and capable of adjusting our priorities, discarding our expectations and adopting new practices which allow us to continue to find ways to live, love, laugh and survive.

above Sand. Perhaps our most sought after, yet least respected building material.

RECIPES

BOTTLE BRICKS

If there is one thing I hope you will remember from this chapter, it is that you are allowed to find, harvest, discover and repurpose your own building materials. You are allowed to be innovative, and I encourage you to be. There is no such thing as waste material, only wasted potential. And simply because a thing was designed to serve one purpose does not mean it cannot serve another.

A well-considered bottle wall can be a work of art, a celebration of both the material and the light that will illuminate it, while also repurposing one of our most commonly used household materials. Bottle walls can be built with many types of mortar, depending on their setting.

While they serve as functional walls, I usually suggest planning bottle walls into the 'finishes' part of your building project. This is because building a good bottle wall requires attention to detail and a zen-like approach. Play some music, take your time and enjoy the process as you work.

Ingredients

- Assorted bottles. Preferably similar sizes and in a variety of colours and shapes. Each bottle brick uses two matching bottles. A square metre of wall will require approximately 90 bottle bricks, so 180 bottles
- Tile-cutting wet saw
- Buckets
- Water
- A brush for cleaning
- Strong masking tape or gaffer tape

Safety note: When cutting bottles on the tile saw, it is very important to wear eye protection, ear protection and heavy gloves in case of sharp glass. I also recommend wearing a waterproof jacket for this activity because there is always quite a lot of spray from the tile saw.

Method

1 The first step is to organise your work flow. You will be cutting the bottles on the tile saw, then washing the bottles, then drying the bottles, and finally taping two dry bottle ends together before safely storing your new building blocks. If you set up your workstation to accommodate these steps in that order, you will find the process will move quite quickly and easily.

2 Set up your tile cutter with a guide so that you can easily cut all your bottles to the same length. A length of 75mm works well, and will create a 150mm bottle brick. Bring the bottle to the blade, slowly touching the glass to it, making sure the butt of the bottle is tight to the guide bar. Now slowly rotate the bottle, allowing the blade to cut evenly through the glass.

3 Once you have cut through the bottle, put the bottlenecks in a container and put the bottle ends at your washing station. When working on my own, I like to work in batches so that I get to change tasks often; I cut around 50 bottles at a time and then take them through the rest of the process.

4 Thoroughly wash the bottle ends, using a scrubber on a stick if necessary. You will see everything through them once they are in the wall and there is no going back then! Set them to dry on a rack, or dry them with a cloth if you are in a hurry.

5 Finally, choose two bottles of a similar diameter and tape the two open ends together, ensuring they are held together tightly.

6 Place your finished bottle brick in a box and repeat the process until you are ready to make your bottle wall.

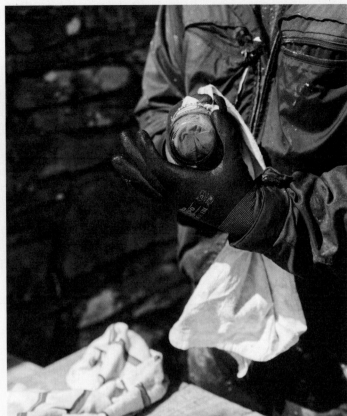

BOTTLE AND CAN WALLS

Bottle and can walls are a great example of building everything out of anything. They illustrate how a material, in this case mortar, can be made stronger by simply altering its shape. The honeycomb pattern in the walls uses less material yet becomes stronger, resulting in excellent load distribution. You could remove the bottles and cans from your wall at the end, and it would be just as strong.

There are few simple techniques that will help you build these walls with ease and confidence.

Ingredients

– Cement mortar, lime mortar or cob mortar, whichever suits your situation best
– Aluminium drink cans or bottle bricks (see page 98)
– Spirit levels, 1200mm and 200mm
– Cloth for cleaning
– Sponge

Safety note: Use rubber gloves and eye protection if you are using cement or lime mortar.

Method

1 Start by making a course of mortar patties 10cm wide by 10cm tall by 15cm long. Lay them along your wall line like fallen dominos, each patty resting on the previous one.

2 If working with bottle bricks, take a brick and place it with its long side perpendicular to the long edge of the patty. Push it down into the mix, ensuring there is an even bed of mortar below. Space the bottle bricks approximately 3cm (two finger widths) apart as you lay your first course.

3 Gently add the next layer of mortar on the top of the bottles, pressing lightly to encourage it to move down between the bottles below. Use a level if you are laying horizontal lines. Also use a small level to ensure each bottle is level to itself as well. These regular checks will keep your wall looking tidy.

4 If you are working with cans, it helps to crush the can slightly; pressing a crease along the side of the can helps it grip the mortar. Keep the ends of the cans two fingers apart and build in just the same way as the bottle wall.

5 Depending on the mortar you are using, you will likely need to stop building after 5–6 courses to allow it time to cure overnight. Adding too much weight onto wet mortar can cause it to spread out at the bottom.

6 Wipe the ends of your bottles with a damp cloth to clean them. It is much easier to clean wet mortar than dry mortar!

7 Next day, the mortar will have cured adequately overnight and you should now be able to groom your wall, filling in the gaps between the cans or bottles. Using gloves, you can apply mortar into the gaps with your hands – wearing gloves! – or with a small trowel.

8 After an hour or two it should have cured enough for you to 'float' the surface with a damp sponge. Floating is the process of brushing a wet sponge over the surface of the mortar. This creates a soft yet grainy level finish.

9 Clean your bottles with a damp sponge followed by a dry cloth, to remove any residue.

TIMBER FLOOR STRUCTURE

Hopefully by now we understand that a material's strength is determined not only by what it is made of, but also by how we choose to use it. Understanding the potential strength of timber is the first step on the path to using this renewable resource. Building our homes up off the ground is standard practice in the majority of the world's wet or humid countries. It protects us from rising damp and rainwater run-off, as well making services and utilities easy to access and repair.

There are several ways to build a structural timber floor. For the purpose of this study we will illustrate just one method, but remember, all the concepts are the same. It is all about gravity and load disbursement. Considering the loads which will impact your structure is necessary before you begin constructing. Please refer to the span chart information on page 88 before you begin planning your floor.

Makes a 4 metre × 4 metre floor

Ingredients

- 14 lengths of pressure-treated timber 150mm × 50mm × 4m
- 100 × 90mm framing nails
- Exterior wood glue
- 3 × 200mm clamps

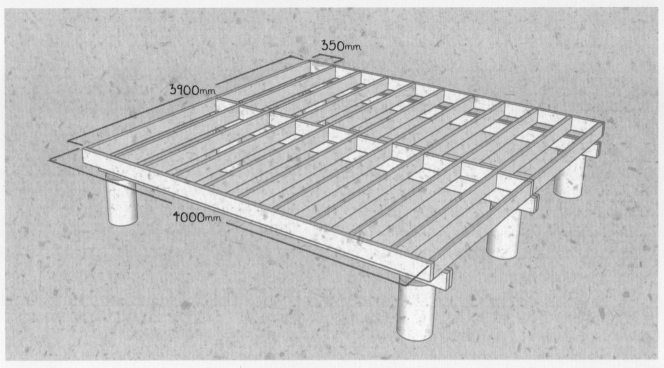

350mm

3900mm

4000mm

Assemblage of a
suspended floor
structure.

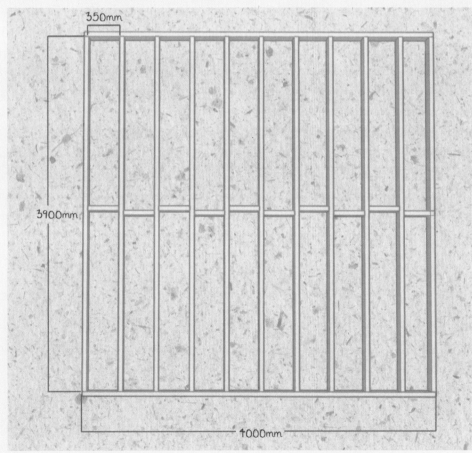

350mm

3900mm

4000mm

Method

1 If you have already poured concrete foundations and did not allow for any way to attach a timber frame, it is not too late. You can make the transition from masonry to timber by drilling a hole through both the timber and the concrete and installing an anchor bolt. As you tighten the nut, the sleeve around the bolt will expand and grip within the hole.

2 You are now going to create three laminated beams. Lamination is the process of joining timber together to create a thicker piece of timber. Cut six lengths of timber, each 4 metres long. Lay two of your pieces of timber beside each other, and apply your exterior wood glue to the flat 150mm-wide surface of one of the pieces. Be generous with the glue, but don't add so much that it spills out when you press the two pieces of wood together. Place the other piece of timber flat onto the glued piece, pressing them together firmly. Line up the two pieces of wood accurately, and fix one nail in each end so that they are securely attached. Stand this piece of combined wood on its long edge, and apply the clamps evenly along the length of your wood to press it together. Add nails, in a zigzag pattern, at 30cm intervals, the whole way along. You can drive the nails in at a slight angle to ensure a very tight grip. You can now remove the clamps and let the nails hold your laminated beam together until the glue dries. Repeat the process with the other four pieces of wood to create three beams which are 150mm by 100mm by 4 metres long.

3 You have options when it comes to transitioning from your masonry work in the foundations to your carpentry work. You can make or buy steel brackets which you can anchor into your foundations using concrete anchor bolts. These saddles support your timber and raise it away from any potential damp. Or you can anchor a pressure-treated timber plate directly to the foundation concrete and use ordinary timber screws to attach your beam. Place your laminated beams across your foundations and attach them by whichever method you have chosen.

4 Now cut two more lengths of timber 4000mm long and cut 11 lengths at 3900mm. The shorter lengths will fit between the longer lengths, and you will end up with a 4-metre × 4-metre square.

5 Mark the two 4-metre lengths at 400mm centres* and mark an X on the side of the line where your timber is to be fixed. Mark both pieces identically. (*Centre positioning, also known as edge-to-edge positioning, means that the distance between the studs is calculated from the centre of one stud, to the centre of the next stud, as opposed to the gap in between them.)

6 You can now lay all your timber across your laminated beams. Do this so that the 3900mm lengths run across all three laminated beams, and the 4000mm lengths fix to the ends and hold them all in place.

7 You can now nail your structure together, making sure to line up your marks and crosses with your timber, always making sure you are on the correct side of the line.

CHAPTER 3

STRONG SHAPES AND WALL STRUCTURES: HOW TO MAKE A STRONG MATERIAL STRONGER

In early 2001, an earthquake tore through the city of Bhuj in Gujarat, India. It lasted only 22 seconds, but it killed almost 20,000 people and destroyed 340,000 buildings. Over 300km away, in the city of Ahmedabad, more than 50 multi-storey concrete buildings were brought to the ground in a pile of dust. But in the small rural villages that surrounded the epicentre of the destruction, the traditional circular earthen huts survived unscathed. Some of these villages were as close as 20km from Bhuj, and yet they remained completely intact. These huts were built with a basic earthen cob mix, which is three parts sand to one part clay, combined with one part loose straw.

The cob is pressed by hand into the wall, knitting each subsequent layer to the last. Round timber lintels over doorways and windows are common, and a round wood and thatch roof keeps the water out. Comparing this traditional construction technique with that of the collapsed structures, which were built using some of our most contemporary construction methods, raises a serious question regarding our current building approaches. How is it possible that these extremely strong materials crumbled when the clay and sand held together? The answer to this question lies in a very simple truth: the strength of a material alone is not enough to resist a force. We must also use the materials in the right shapes. In the case of the 2001 earthquake that shook Gujarat, it was the shape of the earthen huts that saved them, rather than the strength of the materials from which they were built.

opposite Mud hut in Kutch, Gujarat, India.

As I wandered the streets of the city of Bhuj almost 10 years after the earthquake, I could still see many scars of the destruction it had caused. Earthquakes, like all forces of nature, will always take the path of least resistance. When an earthquake, travelling laterally across a landscape, encounters a flat concrete wall it has two options. It can try to turn at right angles, spread to the corners of the building and then travel out the other side back into the ground, or it can break right through it. We saw in Chapter 1 that all forces act in similar ways. The wind, gravity, a running river and an earthquake are all forces of various magnitudes that move through the landscape, meeting resistance as they move. When that happens they will either break through that resistance or go around it. Like the wind catching the sail of a ship, or the flat surface of your paddle when you are kayaking across a lake, the flat concrete walls of Bhuj offered too much resistance for the earthquake to move around them. It was less effort for the force to break down the wall than to change direction and go around it.

When the same earthquake reached the circular earthen village huts, it altered direction to flow around them and reconnected on the other side, like water flowing around a pebble in a river. Instead of challenging the force in a battle of strength, the circle allows the force in and gives it a path out.

right The combination of the shape of the building and the flexibility of the material makes the huts highly earthquake-resistant.

opposite Tall circular structure in Santa Fe, New Mexico.

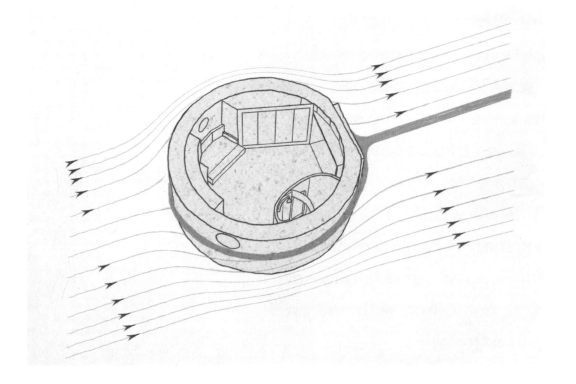

Circles, arches and curves

We have been building circles, arches and curves since humans first started stacking stones and pressing clay with sand to make cob. When we consider their properties, we can see the logic in building with these shapes, particularly when using massive materials such as stone, earth and concrete. This is because a well-balanced arch will only get stronger when more force is applied to it. The path that the load needs to travel along becomes clearer as the connection points compress and tighten.

We relied exclusively on compression when first building stone arches. That is to say, we were actually relying on gravity and using it to our advantage, rather than resisting it. The limestone bridge connecting Christchurch Cathedral with the Synod Hall in Dublin's city centre is a beautiful example of this type of construction. The stones of the arch are cut precisely so as to wedge tightly into each other, allowing the force of gravity to reinforce its strength.

top Load path through an arch with buttresses.

bottom Load path through an arch with flying buttresses.

opposite Load path through an arch.

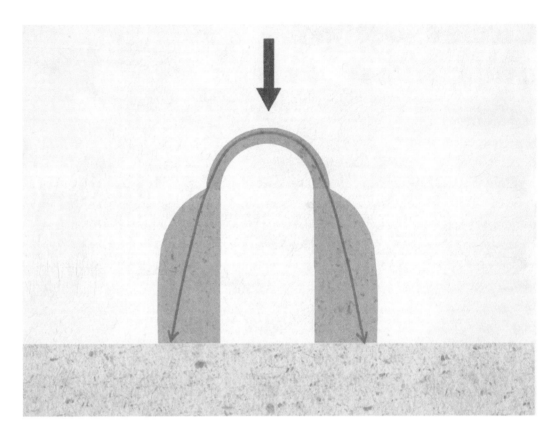

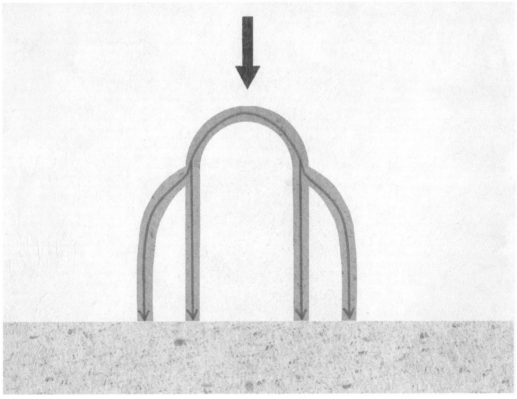

To understand why shapes are so vitally important in the evolution of building, we must learn to observe and predict the path a load will take when it encounters a structure. You can see from the illustration below how an arch receives a load and distributes it down to the ground. As gravity forces all the components of the arch into each other at the same time, it removes the need for adhesives or mortars. The greater the load, the stronger the arch will become, as the bricks are forced together. As the force follows the load path down either side of the arch, it reaches a point where it may find it easier to head off sideways rather than continue to the ground. To compensate for this potential diversion, we learned to add buttresses to our arches. These buttresses are carefully positioned to give the load a direct path to the ground. The innermost part of the buttress is technically superfluous, and can be removed without affecting the strength of the structure. This design, known as a flying buttress, became popular during the Gothic era of architecture.

Wonderful examples of ancient arches and vaults can be seen all over the world, still holding strong after centuries. They show an intrinsic and instinctual understanding of the physics of structures, regardless of whether these concepts were widely understood or taught back then. It is most likely that the successes and failures of these builders of the past, and their mastery over the forces of gravity, were attributed to the work of the gods.

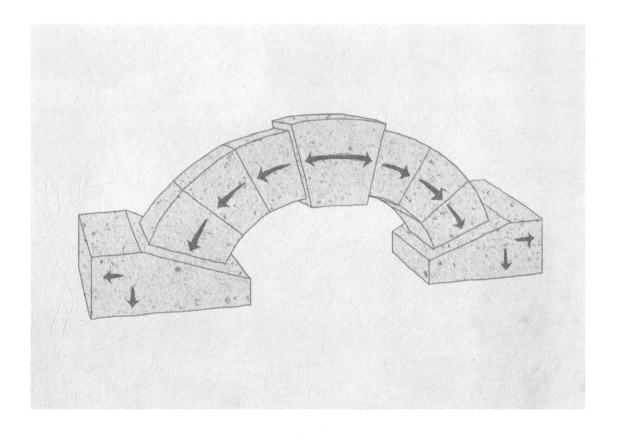

An arch is a section of a circle or an oval. An arch in a structure is able to divert the flow of gravity back down to the ground, in the same way that the circular huts of the Kutch Desert were able to divert the flow of the earthquake. A triangle will do the same thing by taking the load and spreading it down and outwards over a wider area. Compression occurs when a material is pushed together between two forces, and it often results in a material bowing or bending. Tension occurs when a material is pulled apart by two forces. Flexible materials are better at handling tension, while rigid materials are stronger under compression. You can see in the diagram that the elements of a triangle under compression are supporting those in tension, and vice versa.

When it comes to building, squares and rectangles are not particularly strong shapes. They have a hard time dispersing loads and are prone to something known as racking. Racking is when a square opens up at its corners, and skews or collapses to one side. We can make a square strong by turning it into two triangles, which will not be susceptible to racking, as they are essentially rigid. Some of our most common sheet materials, such as plywoods and oriented strand board (OSB) are strong because of the triangles they are made up of. The way we use triangles to create strong walls can be understood by considering the load path again. If we look at the diagram to the right, we can see how the load is distributed evenly throughout the stud wall, and down to the ground, via the foundations. To understand how best to brace a wall, we must first understand where the load path is going to travel, and how best to accommodate that. We then add diagonal bracing to the wall, turning its rectangular shape into two triangles.

opposite Arches beyond arches in Co. Clare.

below Load path through a triangle. Note the simultaneous compression and tension.

bottom The effects of compression and tension under a lateral load.

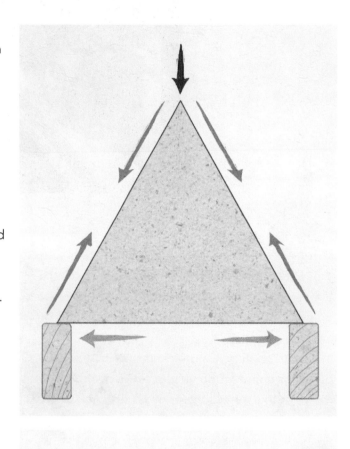

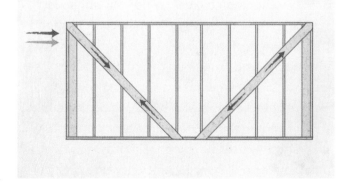

Diagonally connecting our furthest points ensures that our rectangle will not be subject to racking. Subsequently, keeping the square rigid also ensures an even distribution of the load down to the ground.

In the 21st century, we are accustomed to our buildings being made up of a series of straight lines, rather than curves, spheres or partial circles. As we have industrialised our material manufacturing, we have moved towards straight lines and right angles rather than the curves naturally provided by nature. Instead of focusing on the power of strong shapes, we started to invest in the wonder of strong materials. The proliferation of structural steel girder, glue laminated timbers and spanning concrete beams are all examples of simply using more of a given material to make it strong, as opposed to being able to use less material in a stronger shape. A combination of storage, efficiency and transport

compounded such architectural evolutions, and the fashion of designing with circles and arches moved further into the background, becoming associated with heritage buildings and ancient churches.

This evolution in building is not wrong, and engineers' reports have proved that the strength capabilities of these materials are up to the job they have been given. But the inquisitive builder in me looks at the earthen huts of Kutch, standing alongside the rubble of their concrete counterparts, and has to question if we are taking all eventualities into consideration. Are the codes and laws that deem materials 'safe' or 'unsafe' encouraging us along a path of discovery and evolutionary success? Or are they creating a monoculture of modern structures, which may or may not be standing in 200 years' time?

Exercise
NOTICING STRONG SHAPES

Look around you. Whether you are in your home, your office, your school or in the street, anywhere you see buildings you will see strong shapes. If you are in a town, look for a bridge to see the art of strong shapes in action. A bridge is usually made up of a beautiful combination of arches and triangles, perfectly balanced and evenly distributing the load, not only of its own weight, but of all the weight it might carry to the ground.

above Spiralling brick column in Co. Clare. The spiral gives the load a direct arched path to the foundation.

right Arched brick outdoor oven in Co. Clare.

The evolution
of walls

We must remember that the strength of a wall is only one consideration. We have known how to build strong walls for a long time. The more pressing matter is how to build structurally sound walls that are also able to keep us warm.

The evolution of wall structure has been consistently informed by our need to protect ourselves from the elements. Our traditional wall materials had proven to work well from a structural perspective, but we continued to search for ways to stop heat loss through our walls. We created denser insulation and started building timber frame stud walls, instead of internal masonry walls, to allow room for our building's services and systems and all of their necessary pipes and conduits. This design was also more thermally efficient. The majority of our roof structures had been made out of wood, so it made sense when we started attaching this to wooden internal walls.

Despite the development of the timber frame stud wall, many houses are still built with concrete block façades, sometimes with no structural purpose, simply so they look the part. Given the undeniable negative environmental impact of mining irreplaceable sand and stone to create concrete blocks, their use in wall structures cannot be justified by loose arguments such as 'Concrete built is better built', or 'It's the right stuff for the job.'

I have thought long and hard on this topic, and have concluded that the justification for the use of concrete blocks in walls can only be cultural. The heavy stone homes built by our ancestors makes me wonder if it is merely cultural heritage that informs our contemporary design process. This is what we are used to in western Europe; we like our homes to appear heavy and dense – it gives us a sense of security that our investment will be safe for a long time.

In the face of the facts, a transition to timber frame building over the next two decades is inevitable. As pressure mounts from all sides, the mining industry will soon have to justify itself, and the timber framing industry seems to have an answer for every problem. Timber is renewable, transport costs are lower, energy consumption during construction is lower, its thermal resistance is higher and its overall cost is lower. By gluing and vacuum-sealing several pieces of wood together, we are now able to make structural beams far larger than anything we could have grown, and they manage to outperform their steel counterparts in both strength and fire resistance. Yes, you read that right! The timber alternative to the rolled steel joists we are so accustomed to using maintains its integrity in fire long after its steel counterpart has melted. This is made possible by vacuuming all the air from the timber laminated beam and replacing it with a fire-retardant glue. The timber will char on the outside, but as there is no oxygen in the beam, the fire struggles to penetrate beyond the surface.

Common wall materials

Let's look at some of the most common approaches to wall construction today (in addition to the recipes from Chapter 1 and Chapter 2).

CONCRETE BLOCK OR BRICK CAVITY WALLS

Both of these wall types are built with an external and internal leaf of masonry, with a layer of insulation embedded between them. The external surface is usually plastered with smooth cement plaster, or left raw, as in the case of red bricks. The internal surface is usually covered with plasterboard.

TIMBER STUD WALLS

Timber stud walls are built using structural-grade timber, up to 225mm wide × 50mm thick. The studs are usually placed at 400mm apart, or at any divisible measurement of 1200, which is the measurement of a standard sheet of plasterboard. The gaps between the studs are filled with insulation. The external surface is sometimes clad with weather-resistant timber, or plastered to resemble a block-built structure. The internal surface is usually made up of plasterboard, but more adventurous finishes such as lime plaster, cork, tile or wood panelling are all options too.

STRAW BALE

There are two ways to build with straw bales. One is to build a timber frame structure with straw bale infilled between the timber elements, and the other is to build a structural straw bale house. In either case, the primary material is insulative straw. The external and internal surface is usually plastered with lime or adobe, to ensure breathability.

COB

Cob walls are built with a mixture of clay, sand and straw, which are combined in a mixing machine or underfoot. The mixture is then compacted and pressed into the walls. The external and internal surface is most often plastered with lime or adobe, to ensure breathability.

RETAINING WALLS

Gabion cages filled with stone or tyres rammed with earth are built into an earth embankment to create wall structures. Walls are usually around a metre thick. The external surfaces are covered with earth, while the internal walls are usually plastered with earth, lime or cement.

House design: one size never fits all

Planning departments are attempting to define a one-size-fits-all solution for our housing needs. So why is it that their regulations and the resulting cookie-cutter homes, so carefully designed to suit the average person, might not suit us at all? Or a material decision that is assumed to meet everyone's needs could actually result in meeting nobody's needs?

The flawed concept of designing for the average human was demonstrated by scientist Gilbert Daniels in a 1950 study of US Air Force pilots. The Air Force enlisted Daniels and a team of researchers to establish why so many non-combat plane crashes and incidents were occurring. Despite the advances and innovations in plane design, pilots were losing control of their planes; on one single day there were 17 separate plane crashes. The planes' cockpits were thoroughly tested, and it was concluded that the majority of the accidents were not due to mechanical error. So Daniels ran a study of over four thousand pilots, measuring their height, weight, leg and arm lengths at ten different points. Until this point it had been assumed that a cockpit designed to suit the 'average' pilot would suit the body dimensions of most of the men (and they were all men) who flew the planes. That average had been established in the 1920s and had not been revisited since. Through this study, it was discovered that of the 4063 pilots Daniels measured, not a single one matched the average on all ten measurements. The fighter plane cockpit that had been designed to suit everyone actually suited no one. The addition of adjustable seats and several other adjustable components immediately solved the Air Force's problem, and undoubtedly saved countless lives.

This particular story has always stood out to me because my grandfather was an RAF fighter pilot, and had written volumes on the excitement and, indeed, danger of being in the air at that time. Something as simple as an adjustable seat could have dictated whether my grandfather lived or died, and whether my father, myself and my daughter would have even be born.

In the context of building, we must consider in what ways are we conforming and adapting our approaches to suit the average now? Insurance premiums, mortgage qualifications, the size of our homes or gardens is determined by comparisons to a predefined average. It is strange to think that we design so much of our world in this way when so few people, if any, actually fit the average.

If you have decided to build or renovate your own home you have already stepped to one side of today's normality. Over the last 40 years, the price of homes has risen so fast that they have become commodities. Borrowing money is the standard approach most people must face to get their foot on the homeowners' ladder. We have to consider the resale value of our homes while we are designing them, which encourages us to make decisions that will appeal to the average person. And as we saw above, this is a challenging prospect with little likelihood of success. Such design considerations can be limiting and indeed debilitating when we are dreaming up how our homes should appear and function.

A unique home design – the Longhouse in Daylesford, Australia.

The obsession with safe and secure investment has created a generation of home design that often lacks culture and purpose. We feel nostalgia and connection with the buildings of our grandparents, the architectural shadows of the way of life we no longer live, but appreciate coming from. Think of 18th-century design and architecture, where every object, from a hair clip to a home, was designed to be beautiful and long-lasting.

You might only build one home in your life, and you should celebrate all the decisions, research and sweat that have gone into the process. Here's to empowering our creative minds, and building thought-provoking structures, not just for the future, but also for the present. My hope is that this book encourages the part of you that wants to live in joy rather than die in wealth.

opposite Spaces within spaces at the Longhouse, Daylesford Australia

above Bespoke masonry gutter on an Earthship in Argentina.

over Timber greenhouse in an Earthship in Argentina.

Understanding pressure-treated wood

Wood is an incredibly sustainable and versatile material. As with all construction materials, it is vital to understand its weakness as well as its strength, so that you can use it to its greatest potential.

Our most commonly used construction timber is pine. It is fast growing and incredibly straight, which makes it ideal for milling into the square-edged timber we are so used to seeing on building sites and in our homes. Pine varieties have many different qualities, and they share some common weaknesses. The rapid speed at which pine trees grow is also why pine is susceptible to deterioration as a construction material. Its speedy growth results in a long and stretched-out cell structure, similar to a series of long tunnels, which transport nutrients and energy from the roots to the leaves and back down again. These nutrient highways in the tree are like voids, giving a gap-filled structure, much like honeycomb. These voids within the trunk and branches are the reason why pine is so lightweight and easy to cut. They are also why it is susceptible to deterioration when used as a construction material.

Small insects and pests think pine is simply delicious, and they can easily burrow and chew their way through the open cell structure. Long-term exposure to water will eventually deteriorate the wood's fibres and its strength. Our solution to these weak points of pine is to pressure-treat it.

Pressure-treated lumber is submerged in a liquid preservative and then placed in a pressure tank, which uses a vacuum to force the treatment to the core of the wood. This process makes the wood able to withstand moisture and also works to deter insects from eating the wood. Unlike painting the treatment on with a brush, the pressure tank ensures that all parts of the timber, right to the core, are treated. The most commonly used preservatives today are copper-based, and can do damage to the respiratory system if breathed in. If you have ever got a splinter from preserved wood you will know it is disproportionately more painful than a splinter from untreated wood. This is due to the copper impregnated in the grain.

For all the advantages pressure-treated wood provides us, it is less than ideal that we are turning an organic, biodegradable timber into a chemical-filled, toxic material which we have yet to find a suitable way to dispose of. This is something to consider when working out which method of wood preservation best suits our projects, and our priorities.

A variety of bio treatments are available that are less toxic, and usually less effective, than pressure-treating. Using specific wood types for specific settings is a solution of sorts, one which requires a variety of wood to be grown and available to us. Yakisugi is a traditional Japanese technique of preserving wood using charcoal and linseed oil to deter insects and prevent water damage. A more contemporary approach is thermowood treatment, a technique in which softwoods such as pine are heated to relatively low temperatures of 180–200°C in a kiln for up to 96 hours. The process causes chemical and structural changes in the wood, and makes it rot-resistant and unappealing to wood-eating insects. Both of these bio treatments create weather- and insect-resistant products without using harmful chemicals or creating a waste product which is difficult to deal with.

To my mind, pressure-treated wood does have a place in modern construction. However, like many of our man-made material approaches, I feel it should be used only when necessary, to avoid creating excessive amounts of material which are harmful to dispose of at the end of their life cycle. Instead of relying so heavily on pine as a cheap and fast-growing construction material, we could be putting energy into planting and nurturing a diversity of construction timbers.

RECIPES

MAKING COB

Cob has been used in construction for millennia. As our original wall structure it is the great-grandmother of concrete and lime mixes, and is the natural point to begin understanding the principles of making and using them. In a cob mix, the clay acts as the binder, the sand acts as the aggregate and the straw acts as the reinforcement fibres, which encourage any potential load to divide and take a different path. There are countless ways to make cob, and I'd like to share the basic method that I have found to be the easiest and most successful. I wholly encourage you to be free and flexible, and to adapt and change this recipe as much as you like!

First you will need to source your ingredients and perform a soil test, which will help to determine the sand-to-clay ratio of your soil. (See the soil test method on page 40.) To create the perfect cob mix, you need enough clay to make it flexible and cohesive, and enough sand to ensure it doesn't shrink and crack when it dries. There is no fixed ratio for this, as the soil is different on every site. The best way to work out the ideal ratio is to do a few test bricks before you begin in earnest. As you do these tests, if you find your cob is cracking, it is likely you have too much clay in your mix. If you find it is falling apart, it is likely you have too much sand. Test out a mix which is one part sand and one part clay, then another which is one part sand and two parts clay, and finally a third mix which is one part sand to three parts clay. This will help you establish what is going to work best for your site.

Ingredients

- One wheelbarrowful of earth. Try to dig down past the topsoil until you hit the sandy clay subsoil, which will work best
- Large barrel/tub
- Cement mixer/hand mixer
- 10–15mm mesh sieve
- Plastic sheet/tarpaulin measuring at least 4m × 4m
- Access to coarse sand
- Straw

Method

1 Dump your wheelbarrow of earth into your tub and add enough water to cover it. Using your hands, or your bare feet, break up the earth, loosening and separating it from any sand and stones. You can either process the earth in small batches in a cement mixer and then pour it into a barrel, or you can use a hand mixer. Soak the earth for as long as you can and then blend it thoroughly, to get the clay and sand to separate, and the sand to be released into the water. If you are not in a hurry, time will do 90% of the work for you.

2 The next step is to push the mixture through a 10–15mm mesh sieve, to remove any more large stones, and the majority of the organic matter, such as grasses and roots, to create a smooth strained clay mixture. You can make your own sieve by stretching 10–15mm mesh over a wooden frame.

3 Spread your mixture across the plastic sheet and add your additional ingredients – sand or clay – and begin blending it together. The consistency you are aiming for is stiff mashed potatoes. I find using bare feet the best method for this, as you can easily feel which parts of the mix need more work. Periodically pulling a corner of the plastic towards you will turn the mixture over on itself, blending it further. If you are determined to stay clean, you can use a cement mixer for this stage, but I would highly recommend getting involved and embracing the mud!

4 Once you feel everything is well combined, add an armful of straw and mix for a further five minutes, rolling the sheet to fully integrate the straw. Your cob mix is ready for use and can be used immediately, or stored for future use.

BUILDING
COB WALLS

Cob walls are thermally massive walls that are sculpted by hand. It is common across the Americas and in many African countries to make cob bricks in small wooden forms. A good size for a cob brick would be 150mm × 250mm × 100mm.

There is no doubt that making cob walls is a labour-intensive process; however, the benefits of these walls far outweigh the effort required to construct them. The raw materials used are either very cheap or free, and can usually be found on your land or locally, offering an extremely low-carbon-footprint wall solution.

Ingredients
– Cob mixture (see page 131)

Method

1 Always ensure you are building onto solid foundations (see page 23) when using cob – it is a heavy material and requires solid footings. Take two handfuls of your cob mix, equivalent to the size of a loaf of bread, and press it together gently, to form your first brick. You want to work it into a shape, handling it just enough to be able to transport and place it, but not so much that it becomes slick on the outside. You want it to remain rough, so that it grips and connects to the adjoining loaves in the wall. Place your first cob loaf. Now place the second cob loaf, leaning it onto the first, like a line of fallen dominos. Repeat this process, placing the cob loaves so that they connect to both the foundations and their surrounding loaves.

2 Once you have laid a few metres of your wall, go back and gently press and squeeze the laid cob bricks with your hand, in a similar fashion to kneading bread. You do this to knit each brick into all its neighbouring bricks. Keep an eye on the overall shape of your wall to ensure it stays straight, and adjust your technique as necessary.

3 Once all the cob bricks have been kneaded together to form the first section of the wall, repeat the process to create the second course. Remember to knead the second course of bricks into each other and also into the layer below. The aim is to create a wall that is totally connected and monolithic.

4 Depending on the mix, I usually find that two to three cob courses is as much as I can build in a day without compressing the cob at the bottom of the wall under the weight of the material above it. When you think you might be nearing that point, prepare your last course of the day by creating a raised spine of cob along the top of the bricks, pinching it together and into a point between your palms. This will ensure that the first course you lay the next day has a strong shape to attach to.

5 Recommence your bricklaying process the following day and repeat this process until your wall is finished. Then sit back and enjoy the view of your first cob wall.

TIMBER STUD WALLS

As we begin to use less material to create larger spaces, we are entering the world of strong shapes with some momentum. Towering skyscrapers and stretching bridges use the exact same principles of strong shapes, compression and tension as the simple timber stud wall explained here. When it comes to creating a contained space, timber stud walls are usually the fastest and cheapest option. As timber is a renewable resource and the tools required to construct a timber frame wall are minimal, it is a wonderfully beginner-friendly method of wall construction. With small walls, such as the one explained here, it is much easier to construct the wall flat on the ground and lift it up once it is complete. We use nails rather than screws when timber framing because nails have much higher tensile strength, meaning that they will bend, rather than break, if under excessive load.

Makes a wall 4m long × 2.4m high with a doorway and window opening

Ingredients

- 12 lengths of pressure-treated pine, 150mm × 50mm × 4.8m long
- 50 × 90mm nails
- Hammer or nail gun
- Circular saw, mitre saw or handsaw, whichever you have
- 10 metres of 10mm wide galvanised band
- Small box of galvanised clout nails 3mm diameter × 40mm long

Safety note: Always wear gloves and protective eye and ear equipment when using any power tools.

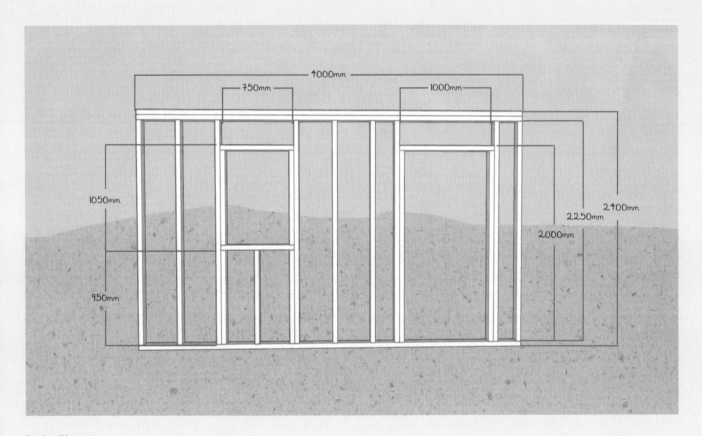

Stud wall layout

Method

1 The first step is to create a cutting list of all the timber lengths needed for your wall. I find it extremely helpful to sketch out a drawing of the planned wall and use this as a reference while creating the cutting list. Our cutting list for this wall is as follows:

– 3 pieces @ 4000mm
– 9 pieces @ 2250mm
– 5 pieces @ 950mm
– 2 pieces @ 750mm
– 1 piece @ 1000mm
– 2 pieces @ 1950mm

2 Use your saw to cut all the lengths on your cutting list.

3 Lay out your wall pieces on the ground, according to your sketch. Note that all horizontal pieces will rest on top of the vertical studs, and not in between. This ensures that the load path continues all the way from the top of the wall to the ground.

4 Whenever possible, your studs should be spaced 400mm apart, on centre. (See page 108 for an explanation of on-centre positioning.) This positioning enables you to make most efficient use of your sheet material, and has the added benefit of allowing us to know where the studs are positioned within the wall in the future.

5 Once you are happy with your layout, it is time to fix it all together. Mark out where you want all your studs, and window and door openings to land. For accuracy, you should mark where you want the edge of your timber to land, and always place an X on the side of the line your stud will go. Use your hammer and nails to attach your wall together.

6 Make sure that all the upright timber pieces are at right angles to the horizontal pieces, so that, when installed, every piece of timber is both plumb (straight up and down) and level (straight from side to side). The easiest method for this is to measure from corner to corner diagonally in each direction and compare the measurements. If they are exactly the same, your wall is square. If the measurements are not the same, adjust the wall by pulling one corner until the diagonal measurements match.

7 Now it is time to brace the wall. You can do this with sheet material, such as plywood, but I prefer to use galvanised banding – it is cheaper and lighter but still does the job. Make a large triangle shape by running the galvanised banding diagonally across the wall. Remember to avoid any door or window openings.

8 Attach the galvanised banding using the clout nails and hammer. Start at one end and attach your banding to each stud behind it with a single nail. To make life easier, you can pull the band towards your end point and apply a temporary nail, hammered only half-way into the wood. Use your foot to press down on the banding to create tension, and hammer the next nail into the next stud.

9 Repeat this process, using your foot to create tension as you apply the next nail, all the way to the end. Now you can remove your temporary nail, and hammer your final nail in position. You should always band in two directions, to ensure racking (skewing) is prevented from both sides.

10 Now get someone to help you lift your wall upright and into position, ensuring that the bottom of your wall is close to its final position before you lift. While one person holds the wall, the other can install two temporary braces, by propping the wall with two pieces of timber laid diagonally against the wall to hold it upright, while you build the next wall.

CHAPTER 4
THERMODYNAMICS: HARNESSING NATURE

During the nomadic twelve years I spent moving, slowly, from Australia to Ireland, I experienced building in dramatically contrasting regions, with a full spectrum of climates. Antarctica is the one continent where I haven't yet tested building approaches in, but I hope one day I'll be lucky enough to find myself there, hammer in hand. From mountainous Ushuaia in Argentina, the southernmost city in the world, to the tropical jungles of Colombia, and from the arid planes of New Mexico's high-altitude deserts to the balmy walnut-growing regions of France, perhaps the most starkly different climates I have experienced are the two countries I have called my home.

opposite View from our home: rain approaches as the sun is setting in the south-west.

Growing up in the hot, dry, east coast of Australia, the days of my childhood were spent scraping melted bitumen from the road surface off the soles of my bare feet. In my adoptive home of the perpetually wet, windy west coast of Ireland, my daughter's childhood memories will likely be filled with wet wool socks inside welly boots, and bringing in wood for the fire. Without realising it, I had moved from a climate which has two to three short months of winter each year, to one that has between two to three months of a very gentle summer.

What struck me over the years I spent as a travelling builder is how nature operates in essentially the same way across all climates. More specifically, the way heat and cool is transferred is the same all around the world. While our house design priorities may be different in contrasting climates, our methods of interacting with nature are similar. The walls keeping my home in Australia cool under the relentless summer sun, also work to keep my home warm in Ireland's winter storms. Whether we were trying to hide from it or harness it, our constant consideration of the sun is the same.

The movement of heat

It is the rapid movement of particles that creates the kinetic energy we know as heat. These particles vibrate on a microscopic level, moving, colliding and sharing their energy with slower, colder moving particles. There are just three ways that heat can move, and understanding them is key to designing your own energy-efficient home.

Radiant heat is the heat we feel when we sit in the sun. It travels through infrared, ultraviolet and radio waves.

right Radiant heat is the way heat travels through infrared waves.

opposite, left Convective heat is the way heat is transferred through a liquid or gas, such as air.

opposite, right Conductive heat is the way heat is transferred through a thermally massive material, such as steel (in this case, the kettle), or stone.

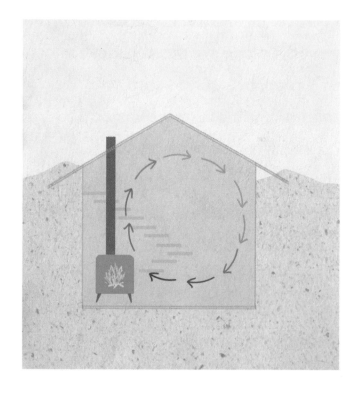

Convective heat is when a fluid or a gas, such as air, mingles with the source of the heat and carries part of that heat away. The heat is actually clinging to tiny water particles in the air. The radiators in our homes, schools and offices, despite their name, actually use convective heat technology to warm our spaces.

Conductive heat occurs when a physical path is created between a warm source and a cold source, allowing the heat to flow freely between them. An example of this is when you put a hot water bottle on your lap and it heats up your skin, or if you are lucky enough to have a willing partner or pet to snuggle up to in a cold bed. The more points of contact you have, the more paths the heat has to travel along. Conductive heat is our most efficient method of transferring temperature.

When a material has the ability to transfer heat well, we refer to it as having a high thermal conductivity, for example an iron stove or a steel frying pan. In contrast, when a material does not transfer heat well, we refer to it as having a low thermal conductivity, for example a pair of woollen gloves, which keep your hands warm by preventing the heat leaving them. The lower a material's thermal conductivity, the higher its insulative value.

Sunlight soaking into
the thermal mass of
brickwork.

Harnessing heat in our homes

Just as there are three ways in which heat can move in a space, there are also three ways in which it can escape a space. In Ireland, we are most often focused on keeping our buildings warm, so we should design our spaces to take advantage of these heat transfer principles. We need to find ways to bring heat in, as efficiently as possible, and simultaneously slow down the heat loss process. So how do we do this? Is it possible to design a home that allows natural warmth in easily, without letting it escape just as easily? The short answer is yes. Through understanding and utilising all three methods of heat transfer, we can maximise the efficiency of our home heating (and cooling), by allowing nature to work for us. Unfortunately, although we can do this, we very often don't.

Now for the long answer.

The contemporary homes which rate most highly on their energy efficiency are built in a similar way to a Thermos flask – an airtight, insulative container with a hard outer shell, into which we pour hot drinks. Our tea will stay warm inside the flask for a few hours, maybe even overnight, but eventually it will cool down. Unless we put more energy into warming the tea again, it will end up being the same temperature as the outside air. In the same way, to keep our highly insulated homes at a comfortable temperature, we need a constant supply of energy. Our means of creating the energy that these homes are heavily reliant on are varied, contentious and polarising. To design homes that run most efficiently and sustainably, we need to harness the power of nature, and use this in parallel with such systems.

Using nature to our advantage

We are surrounded by reliable, perpetual heat sources and we are constantly evolving technology to be able to harness and store it. Ground source heat pumps are a great example of how we can use the laws of nature. They use the same technology as your fridge, freezer and car air-conditioner. They function by running a refrigerant liquid, with a boiling point of around -5°C, through an underground pipe. The soil one metre below the surface in Ireland is around 10°C, whatever the season. Exposed to the temperature of the soil, the liquid changes to a gas, which can then be converted by a compressor to physically compress the gas molecules back into their liquid form. This conversion from a gas to a liquid produces a large amount of heat as a by-product. The heat can then be stored and used to warm our buildings. The compressor does require electricity to run, but it uses substantially less energy than would be required to create heat directly from electricity, such as in electric radiators. Air to water heat pumps use the same technology, but they use air temperature as opposed to ground temperature. They work well in Ireland's mild climate, and are usually cheaper to install than underground systems.

Another wonderfully reliable natural heat source is, of course, the sun. Rising every morning and following a predictable path across the sky, the sun is a perpetual and free source of radiant heat. The challenge we face with radiant heat is catching it and storing it. We know that ultraviolet light and its associated radiant heat can travel through air and glass. We feel it on our skin, and in the warmth of the stones at the beach. The radiant heat which enters

our homes through glass does not directly heat the air through which it is travelling. It must first be absorbed into the materials of our home, our floors, walls and furniture. The absorbed heat is then released into the air of the home via conduction and convection, due to the difference in temperature between the hot material and the cooler air. This is why it is vitally important to consider what materials our homes are built of, and evaluate their abilities to absorb the sun's wonderful, free heat.

opposite Beautiful inclusions of thermal mass in sun-drenched areas.

below The thermal mass of the city streets absorbs the sun's radiant heat.

Thermal mass and insulation

Thermal mass describes dense materials that have the ability to store and transfer thermal energy easily via conduction, and release it slowly via convection. This characteristic can be used to maintain even temperatures within a space, when the outside temperature is fluctuating drastically. The temperature of the mass is easily maintained, and does not change at the same speed as the temperature of the room. All materials have a thermal value, and are measured by how quickly they transmit heat.

Much like charging a battery, it can take time to heat thermal mass, simply because of the huge amount of energy it absorbs. The good news is that it also takes time for the mass to cool down, while it releases all the heat energy back into your home. A city's pavements and brick façades won't feel warm first thing in the morning, but by late afternoon, and, indeed, long after the sun has set, they will radiate the sun's energy back into the streets. When sunlight hits on thermally massive materials such as stone, concrete, tile or earth, the mass will store that heat, conducting it throughout itself and slowly releasing the heat back into the room. These materials are much more efficient at storing heat than insulating materials such as wood, fabric, cork or plasterboard. Acknowledging that the sun is our most abundant source of free heat energy allows us to design our homes in such a way to harness it.

To understand how thermal mass works on a small scale, think how a steel cup full of hot coffee would burn your hands, while a ceramic cup is usually comfortable to hold. The steel is thin, while the ceramic cup is thick. Both materials are highly conductive, but it is this difference in thickness, which equates to a difference in volume, that contributes to the cup's thermal mass. The more thermally massive an object, the more heat it can absorb, before transferring it to the outer walls and, in this case, to your hands. A thick steel cup will also be comfortable to hold, but will heat up faster than a clay cup, because the atoms of steel are so closely pressed together, and this provides easy pathways for the heat to travel quickly. Clay, in comparison, is relatively aerated, and the air gaps in it will slow the heat travel down.

Both the steel and the clay are thermally conductive materials; however, the clay takes longer to heat up, and also to cool down, providing us with a temperate and comfortable level of heat over a sustained period. Now imagine your steel cup has a small handle attached; it will be more comfortable

right How a rug, concrete floor and underfloor insulation interact with radiant heat. The rug is unable to absorb the radiant heat, while the concrete flooring is thermally massive and absorbs the heat, storing it and releasing it back into the room over a sustained period. The underfloor insulation ensures that the heat stored in the concrete floor is forced back into the room, and not into the earth or surrounding building.

to hold by the handle. Why is the rest of the cup hot while the handle stays cool? It is because the physical connection between the handle and the cup provides only a small amount of thermal mass for the heat to conduct through. The majority of the cup's heat is lost through the surface of the cup into the air, via convection. The good news is that for us as builders, the effects of thermal mass are not limited to coffee cups and can be scaled up to the size of a house with the same results.

With more intense heat sources, such as fires and stoves, we often create excess heat energy, because they can create such high temperatures. If we can harness this excess energy and store it in thermal mass, it can then be released back into our buildings, when they become cooler than the mass and allow the mass to expel its stored heat. This would reduce the amount of time and fuel needed to maintain comfortable temperatures in our buildings and allow convection to draw heat

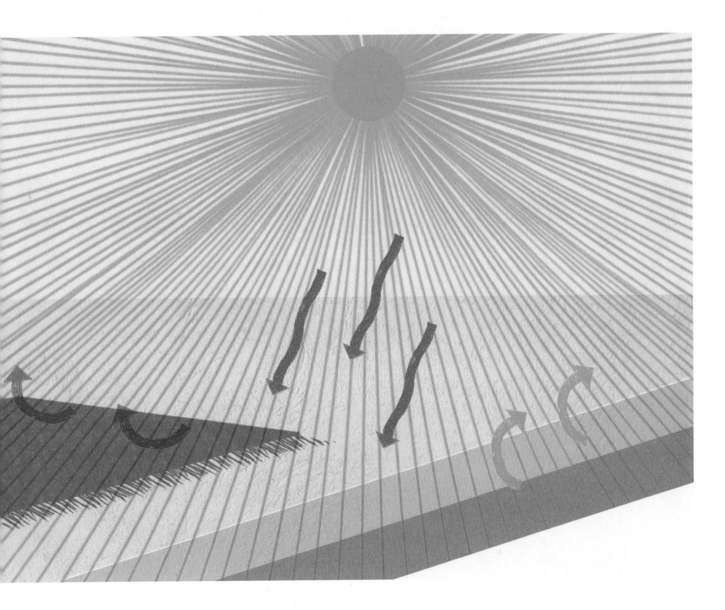

from the thermal mass as required, to equalise the temperature of the space.

Incorporating and dispersing sources of thermal mass throughout our living spaces allows for a more even, comfortable heat. It reduces the hot and cold points that are created by point heating, such as radiators and electric heaters, in a highly insulated home. Underfloor heating is a successful example of a thermal mass heating system, which stores and releases moderate, consistent heat. This is achieved by running a network of hot water pipes within a concrete floor. Conduction transfers the heat from the hot water to the concrete, and convection heats up the air touching the concrete and transfers it to the room and your body.

The original Irish cottage was designed around a huge body of thermal mass, in the form of the stone fireplace and chimney. These were traditionally built from locally sourced stone or, occasionally, from wattle and daub. The fireplace was really the heart of the home, providing warmth, light, a place to cook, tell tales and dry off after a day of working outdoors in the elements. Often a bedroom would be positioned behind the fireplace to allow the residual heat from the thermally massive wall to be enjoyed in the dark hours of night.

Another common example of a thermally massive heat source are masonry stoves, which can be found across Europe and are still very popular in many Scandinavian countries. I was lucky to train in masonry stove construction with stove master Lars Helbro in Denmark a few years ago. I have always enjoyed the unique design of these stoves and the juxtaposition between the utility of common clay bricks and the soft curves that are created

by rounding off the brick's corners. Stove making is a true art form, and not only requires a deep understanding of the material you are working with, and how it will react to intense temperature changes, but also a knowledge of how fire, smoke and heat travel and react to different influences. Like the Irish cottage fireplace, the core concept of the masonry stove is to create and store heat using thermal mass. The key difference between a masonry stove and an iron stove is that a masonry stove can store heat long after the fire has gone out. Iron stoves are good at transferring the heat of the fire quickly to a room, but will cool down quickly once you stop adding fuel. In contrast, bricks and other thermally massive materials have an amazing ability to receive, store and regulate the heat that is passing through them. They will rarely get too hot to touch, and a single leaf of bricks around a stove can stay warm for up to 12 hours, consistently radiating heat into the room.

While stoves and huge stone fireplaces may not be a big part of our clean energy future, the concept of thermal mass should be. Any amount of thermal mass in a home will store the ambient temperature of that home and re-admit it into the house. The fastest way for heat to enter and leave a thermal mass is via conduction or direct touch to a cold surface, or the outside. I believe thermal mass should be, at the very least, concentrated around our heat sources to capture the excess heat which is created and release it back to us slowly, at temperatures we can enjoy.

opposite Solar-oriented greenhouse for plants and people alike.

Thermal bridging

The fact that thermal mass is highly conductive is both its greatest advantage and its greatest weakness. When connected to a cold body of mass, such as the ground, or exposed to the air outside, it will leak its stored energy, and will eventually become cool again. This dilemma is known as thermal bridging. Like a leak in a bathtub, these points around the home present the path of least resistance for the heat to escape, or for the outside air temperature to enter. It is an issue with both conductive and convective heat loss. So you can see the importance of understanding the different ways that heat travels, as it allows us to predict the location of any potential weak points in our buildings, and take measures to prevent thermal bridging.

Insulation

Materials which have a low thermal conductivity are generally referred to as insulating. They insulate one space from another, and allow us some degree of control over the temperature of such spaces. Insulation is not a modern concept, but we have gotten better at creating new insulating materials, through understanding how materials resist the transfer of heat.

If conduction is the most efficient path for heat to take, then creating a clear break in conduction is our best way to slow it down. One example of how we can do this is glass windows. Glass, on its own, is a dense, highly conductive material, which is why single-glazed windows are so cold

right How heat travels through different mediums. This shows the effectiveness of double-glazed glass with a gaseous barrier between the thermally massive glass panes.

opposite Insulation deflecting and dispersing heat. Notice the overall effect of slowing down heat transfer.

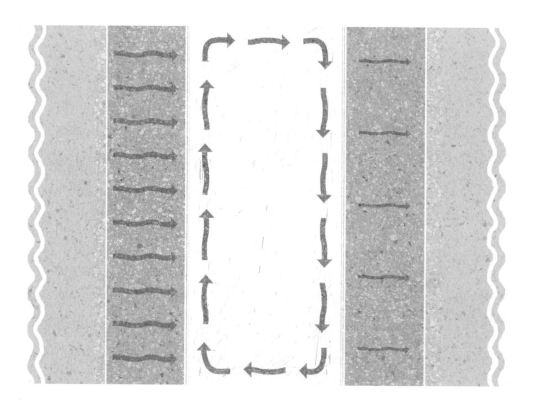

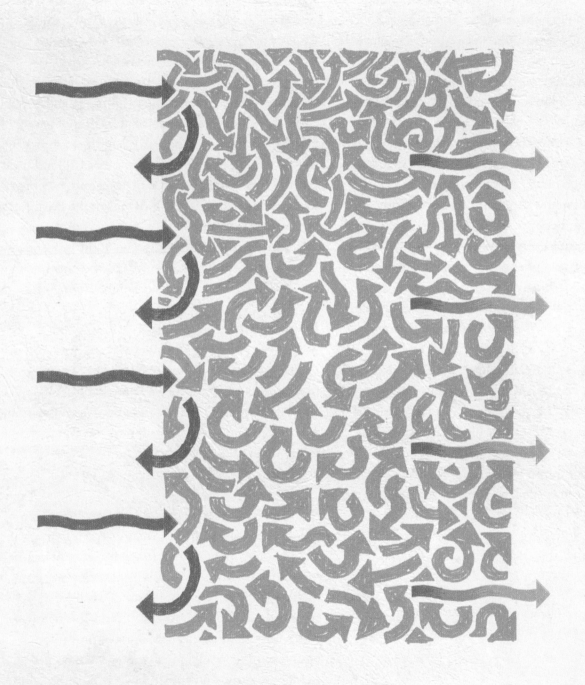

to touch and why heavy curtains were often a necessity in the homes of the past. The invention of the double-glazed window was a game changer when it came to heat retention. The original double-glazed window was made of two panes of glass separated by an air gap. This meant that while the internal glass very quickly adjusted to match the temperature of the room, it could not conduct the heat to the outside. Instead, it had to transfer the heat through convection to warm the exterior glass, slowing the heat loss process down considerably – but not nearly enough to dispose of the curtains yet. The water between the glass panes had a frustrating tendency to collect on the surface of the glass during quick temperature changes – this is the commonly known phenomenon of condensation. The next step in the evolution in windows was to fill the space between the two panes of glass with an inert gas. The gas proved to be much more insulating than the air that previously existed in the cavity, because the water particles suspended in the air would carry the temperature between the two panes of glass. The inert gas substantially slowed the convection process and led to an incredibly efficient design, which allows radiant heat to travel into our rooms via sunlight, while preventing convective and conductive heat escaping through the glass.

opposite A typical cavity wall, showing insulation and wall ties.

THE EVOLUTION OF INSULATION

Before advances in transportation enabled us to move materials long distances, building with what was available in our vicinity was our only option. Coastal areas often provided the materials for stone houses. Inland lakelands often provided clay for adobe and mud houses. Densely forested woodlands enabled us to build timber frame houses. Lime, the predecessor to cement, was another common, locally sourced natural material that was ubiquitous in home construction across Ireland and Europe. It was the demands of cities and their need for construction materials that fuelled the drive for more consistent and easy-to-transport materials. Of our man-made materials, one of the oldest and, indeed, long-lasting must surely be the clay brick, and that is where we can start our journey.

A standard brick townhouse in the early 1800s would have been constructed with common red or yellow bricks, which interconnected between the inner and outer walls. As we know, clay is a highly conductive material, and so the heat from the inside walls of the house was transferred, via conduction, to the outer walls of the house, where it was quickly carried away by the convective air currents.

It became common practice in the early 1900s to build two single-leaf brick walls with a 50mm air cavity between them. The two walls were held together with iron wall ties, which were small iron rods embedded into the mortar of each wall, allowing them to support each other. This air cavity, much like a double-glazed window, prevented heat loss via conduction, and forced the heat to travel, airborne, between the two walls. From what we know about how insulation works, we can understand that air is not a particularly effective insulator. It wasn't until the 1990s that it became common practice to fill the cavity with fibreglass batts or rigid insulation boards. Little has changed in the last 30 years in brick and block construction. The size of the cavity has increased, and the amount of insulation along with it, and wall ties are now made of stainless steel to prevent them rusting. Filling this cavity made a very big difference to the temperature of our spaces and, much like layering your chest with a woollen jumper under a windproof coat, it allows us to reduce the powerful effects of convection against our buildings.

Unfortunately, we often have a tendency to solve problems with interventions that introduce new problems; and these new problems require new solutions; and so on. While we discovered how best to keep cool air out of our buildings, we managed to lose one of the key components of storing and releasing heat. The use of thermal mass in our old homes regulated the internal temperature of our buildings in a way we are now trying so hard to do with costly technology and mechanical intervention – radiators, heat pumps, gas and oil boilers, air exchange systems and air-tightness membranes – the list goes on.

Removing thermal mass from our homes was a short-term solution, applied in an attempt to stop heat escaping our spaces via thermal bridging. I wonder what might have happened if we had gone back in time and applied the knowledge, tools and methods we have developed over the past hundred years to build structures that take advantage of the thermal slowing qualities of insulation and the thermal storage qualities of thermal mass. The great thing about building or renovating a space yourself means that you can harness the power of these concepts without the need for any time travel whatsoever!

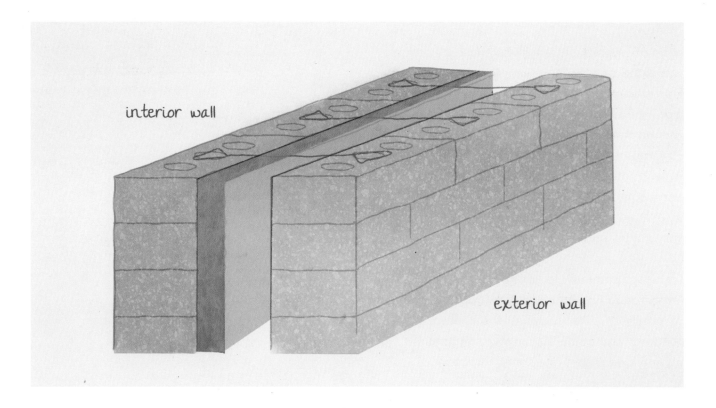

interior wall

exterior wall

Earthships and harnessing nature

Let's examine the design of an Earthship building as an example of these two concepts working in harmony to create a home that self-regulates its internal temperatures without the need for mechanical intervention or any additional fuel-fed heat source. The entire layout of an Earthship is designed to capture and store as much passive energy as possible. One whole façade of the building is clad with two layers of double-glazed glass, which allows radiant solar heat in, without letting the warm air escape. This heat is stored in the thermal mass of the floors and the walls, both of which are wrapped in insulation, to ensure that conductive heat loss does not occur. This heat is then slowly radiated back into the building, regulating the air temperature with the stored heat in the floors and the walls.

I had an opportunity to test the efficiency of this combined system when I was staying in one of the Earthship rental homes in Taos, New Mexico, in a high-altitude desert. It was the middle of the winter, with an outside air temperature of 2°C and an inside air temperature of 16°C. There was over a foot of snow on the ground. I put on my quilted coat, hat and boots and opened all the windows and doors to let the cool night air swoop in, while I went for a starry night walk.

I returned an hour later to find that the indoor temperature had matched the outdoor temperature of a chilly 2°C. After closing all the doors and windows, I sat at the kitchen table and poured myself three fingers of whiskey. I waited.

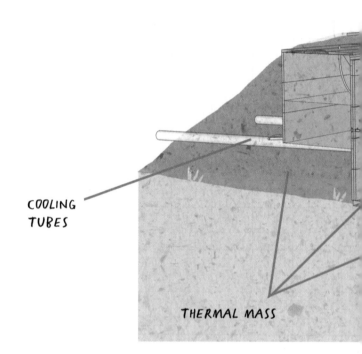

COOLING TUBES

THERMAL MASS

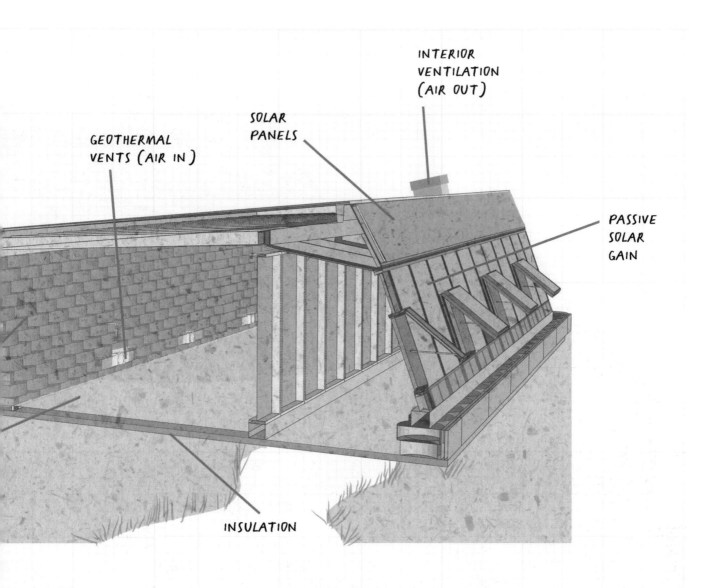

GEOTHERMAL
VENTS (AIR IN)

SOLAR
PANELS

INTERIOR
VENTILATION
(AIR OUT)

PASSIVE
SOLAR
GAIN

INSULATION

Section of an Earthship
building design, a
unique and highly
efficient combination
of thermal mass and
insulation to maximise
the potential of the
sun's radiant heat.

An Earthship home.

After one hour, the air temperature was 4°C. After three hours, it was 8°C, and by the time I woke up in the morning the house had reached 15.5°C. This was all without a drop of fuel or the striking of a match. The heat stored in the building's thermal mass radiated back into the room and heated the cool air. Fast-moving warm air particles displaced the slow-moving cool air particles, which sink down low, and this created a convective flow of warm air.

R-VALUES, K-VALUES AND U-VALUES

I should warn you that this topic can become a little dry and technical. The numbers, values and equations are all too reminiscent of secondary school algebra, and I won't hold it against you if you choose to skip ahead, and revisit this when you feel like stretching your brain a little. However, I will say that understanding how to read and rate our materials according to their thermal abilities can be very useful when sourcing and comparing insulation types.

right The U-value of a wall make-up.

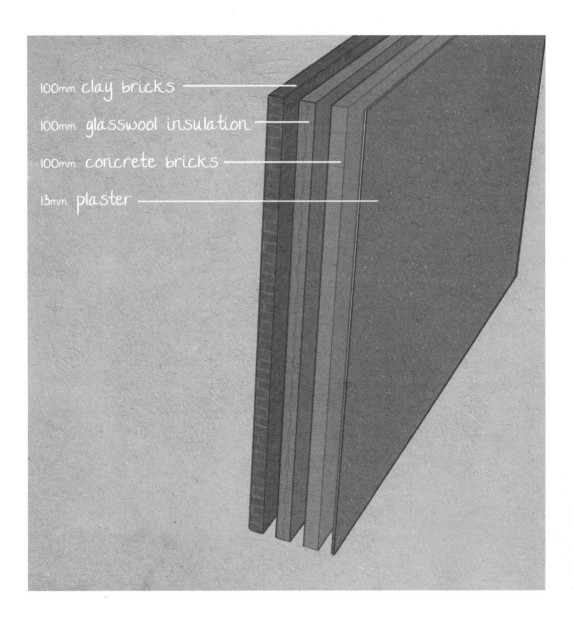

100mm clay bricks

100mm glasswool insulation

100mm concrete bricks

13mm plaster

K-values are the average speed at which temperature will move through a uniform substance. They are also expressed as the time it takes for the temperatures on either side of a substance to equalise.

An **R-value** is the measure of a material's thermal resistance. A higher R-value means a greater thermal resistance. R-values are cumulative, meaning the more you have of a material, the more insulative it is. This can more easily be understood by an example:

Mineral wool has an R-value of 3.0–3.3 per inch. If a wall were to have 6 inches of mineral wool installed in it, it would have a total R-value of 19. Sheep's wool has an R-value of 3.5–3.8 per inch. If a wall were to have 6 inches of mineral wool installed in it, it would have a total R-value of approximately 22.

The R-value of a material is calculated by multiplying the K-value by the distance, or depth, of the material.

U-values are the measure of a material's thermal conductivity, i.e. how easily temperature moves through it. An individual material has a U-value, and a combination of materials used together within a building will have an accumulative U-value. The better insulated a structure is, the lower its U-value will be.

The U-value can be calculated by finding the reciprocal of the sum of the varying thermal resistance of each material making up the surface in question. Unfortunately there is no simpler way of explaining that! Hopefully an illustration will help.

In this illustration we have a wall comprising the following layers: outer layer, 100mm of clay bricks, 100mm of glass wool insulation, 100mm of concrete bricks, 13mm of plaster, plus the internal surface. The U-value calculation would look as follows:

Material	Thickness	Conductivity (K-value)	Resistance = Thickness ÷ conductivity (R-value)
Outside surface	–	–	$0.040 \text{ K m}^2/\text{W}$
Clay bricks	0.100m	$0.77 \text{ W/m}^2\text{K}$	$0.130 \text{ K m}^2/\text{W}$
Glass wool	0.100m	$0.04 \text{ W/m}^2\text{K}$	$2.500 \text{ K m}^2/\text{W}$
Concrete blocks	0.100m	$1.13 \text{ W/m}^2\text{K}$	$0.090 \text{ K m}^2/\text{W}$
Plaster	0.013m	$0.50 \text{ W/m}^2\text{K}$	$0.026 \text{ K m}^2/\text{W}$
Inside surface	–	–	$0.130 \text{ K m}^2/\text{W}$
Total			$2.916 \text{ K m}^2/\text{W}$
U-value =		1 ÷ 2.916 =	$0.343 \text{ W/m}^2\text{K}$

Commonly used insulation types

All materials, whether found, foraged or purchased, have a thermal quality, and it can be helpful to be aware of this, even if you are not using a material specifically for its thermal qualities within your building.

There are a plethora of insulation options to choose from, with varying levels of conductivity. More often than not, your instinct will help guide you on this. When something is heavy and dense, it is usually thermally massive, for example concrete, stone, clay, water or earth. When something is light and aerated, it is usually insulative, for example straw, polystyrene, wool or cork.

A product's thermal conductivity is only one of the deciding factors when it comes to choosing insulation. Other considerations include the overall cost, ecological footprint, ease of application and lifespan of the product. Generally, the more insulative a material is, the more expensive it is likely to be; and oftentimes, the more environmentally friendly a material is, the higher its cost. This, rather frustrating, reality is likely influenced by the labour-intensive production process of many of the more environmentally friendly materials, and the fact that many of them are relatively new to the market. Whichever way you decide to go, you will need to explore your options and weigh up your priorities before you choose the insulation that is right for you.

FIBREGLASS BATTS

These are made by weaving strands of glass and silica fibres together to make a glass blanket of sorts. They are made up of many small fibres laid in opposing directions, with just as many gaps in between. Much like a glass window, the material is not technically insulating, but its composition means that heat has a very hard time moving through it.

RIGID FIBREGLASS BOARD

This is made of the same material as a fibreglass batt, but it has shorter silica strands, as it does not rely on the weave to hold it together. Millions of small strands of glass and silica are sandwiched between two layers of reflective foil, which reflects radiant heat.

MINERAL WOOL

Despite its deceptive name, this insulation material is made in a similar way to fibreglass, from a variety of different materials, including basalt, glass and recycled slag, from the steel industry. The process of creating mineral wools can be compared to making candy floss. After it is crushed and melted, it is poured, as a molten liquid, into a spinner. This creates mineral strands, which can be pressed into convenient bales.

CELLULOSE

Essentially pulped newspaper and cardboard, which is mechanically blown into cavities to create a tight insulation.

POLYSTYRENE

A foamed plastic that you will often have come into contact with in your day-to-day life. Polystyrene is the polymerisation of styrene – basically the plasticising of an oil in benzene.

AEROGEL

A word you will likely hear more frequently in the coming years. This is essentially a gel that has had its liquid removed and replaced with a gas, without damaging its cellular structure. It is currently viewed as the most insulating material on the planet, at 99.8% air, and able to withstand temperatures of over 1000°C with almost no heat transference.

SHEEP'S WOOL

This is an incredible insulator and natural material, and it has been used in dwellings for hundreds of years. Indeed, we still build houses with it in many parts of the world today. To meet fire safety codes, the wool must go through a scouring process, where it is heated several times to remove all of its naturally occurring lanolin oil. The wool is much less likely to catch fire with the lanolin removed, but it is now susceptible to moth and insect infestation, which the oil had previously protected it against. Back to the issue of fixing one problem and creating another! As the scoured wool is combed and layered, it must be treated with a high-alkaline solution to deter insects.

MYCELIUM INSULATION PANELS

Made by inoculating a bed of cardboard or paper pulp with fungi spores. The fungi spreads throughout the growing medium until it has filled the form, creating a dense, honeycomb structure as it travels. Its growth is then arrested by spraying a salt solution over the fungi. The result is a fragile foam material which can be supported on either side by sheets of rigid card.

opposite Commonly used fibreglass batt insulation.

This list of commonly used insulation types is by no means exhaustive, but the characteristic they all have in common is that they are made up of millions of small particles. The smaller the particles, the more insulating the material, because there are more gaps, and subsequent changes of direction, that the heat has to navigate and bypass. The more complex the maze, the longer it will take the heat to escape. This is why understanding thermal conductivity and what makes something high or low in its ability to conduct heat is important in assessing what materials you should insulate your structures with.

Between the writing of this book and its publication, there will be leaps and bounds in the development of insulation materials for our homes and structures. Developers of mycelium-based insulation are working hard to make it as affordable as its unsustainable competitors, and we are not so far away from being able to grow full buildings, rather than build them. Nobody yet knows what our buildings of the future will be made of, though the journey we take to get there will be exciting and, likely, turbulent. As all good journeys are.

opposite Solar relationship orientation cycle as observed from our home at 52°N.

Understanding the path of the sun

While the methods, materials and make-up of house building has evolved drastically in the past few hundred years, our relationship to the sun is one factor that has remained unchanged. The only difference now is that our decisions are based on scientific understanding rather than intuition. This is clear from studying ancient sites such as Stonehenge, or Newgrange in Co. Meath. The magic of this tomb's placement and orientation is revealed at the winter solstice each year, when the light of the rising sun passes through a 19-metre-long passage to flood the inner chamber. Further afield, the Abu Simbel temples in Egypt display a similar understanding of the path of the sun; a doorway directs sunbeams into the inner sanctum and onto statues of King Ramses II and Amun, the god of light. Indeed, it was not just the spiritually or economically elite who were invited to enjoy the luxury of accurate solar orientation of their buildings. Just look at the design and positioning of so many of the dry-stacked stone cottages dotted around the Irish countryside, the majority of which have three or four windows facing south and one facing east. You will very rarely see such cottages with north-facing windows, because there was no passive energy to be gained from this angle.

The science of solar positioning and movement is detailed and mathematical, but for the purpose of establishing the ideal orientation of your buildings we can keep it simple. From our perspective here on Earth, the sun moves around the Equator, rising in the east and setting in the west. Of course, this can be a little discombobulating when moving from the

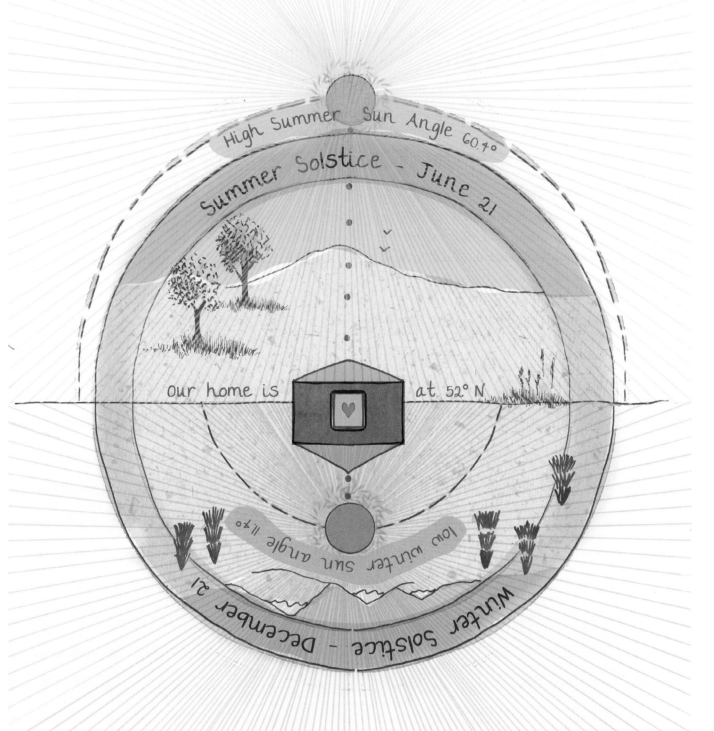

southern to the northern hemisphere, as the sunrise will switch from your right side to your left side. But as long as you think in terms of east and west and not left and right, things should remain clear.

The height of the sun in the sky changes, depending on the time of year and the latitude of your location. The closer you are to the Equator, the less extreme these differences are. In the northern hemisphere, our homes are best oriented to the south. The sun will be positioned close to the horizon during winter, reaching its lowest point on 21 December – the winter solstice. Its position creates a smaller arc, resulting in shorter days and longer periods of darkness. Inversely, the sun reaches its zenith, or highest point in the sky, at the summer solstice, 21 June. This position creates a much bigger arc and gives us many more daylight hours. Of course, in the southern hemisphere the summer solstice occurs on the same day as the northern hemisphere's winter solstice, and vice versa.

Understanding the sun's path across the sky should inform your building design more than any other factor. Designing in a way that harnesses the sun's energy should add no additional materials or costs to your building project. It will only make your house warmer. Our need for heat and light in our buildings is greatest during the winter, so it is important to determine the arc of the midwinter sun and ensure there are no obstructions which could prevent that sunlight coming into your home. Deciduous trees lose their leaves in winter, so they will let the light through, but be wary of evergreen trees, as they are often the culprits for casting large shadows over our buildings.

left Solar angles for summer and winter, taken at our house at 52°N.

opposite Following the movement of the morning sun.

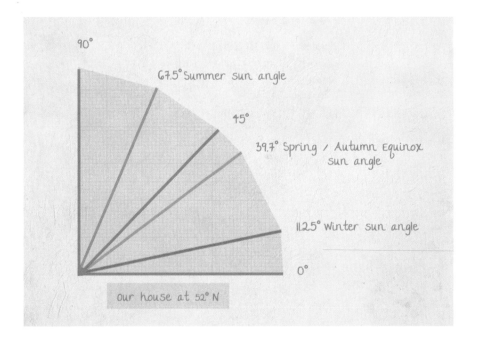

90°

67.5° Summer sun angle

45°

39.7° Spring / Autumn Equinox sun angle

11.25° Winter sun angle

0°

our house at 52° N

Exercise
HERE COMES THE SUN

If you know the latitude of your site, it is relatively easy to work out where the sun will sit in the sky in midwinter. Lines of latitude are the east–west lines around the globe that help us navigate and calculate our seasons. The Equator is at 0° latitude and rises in both north and south directions. The polar circles (the Arctic Circle in the northern hemisphere and the Antarctic Circle in the southern hemisphere) are at 66° latitude. At the Arctic Circle, the midwinter sun rests on the horizon, or at 1° altitude. My home in the west of Ireland is located at 53° north of the Equator. This is 13° less than the Arctic Circle, which means that my midwinter sun sits at 13° above the horizon. If you imagine the horizon is 0° and straight above your head is 90°, you should be able to approximate where 13° lands between those two points.

Once you are able to determine the sun's angle in relation to your house, you can design in a way that maximises the potential solar gain. As well as the size of your building's windows, the roof eaves, gutters and wall height will all affect how much sunlight gets into your home. Working out which rooms will be used most at different times of the day may help to inform your decisions. A snug to drink your morning coffee in would be best positioned on the east side of the house, while your daytime office and kitchen might be best positioned on the south side of your house. An outdoor deck to enjoy sunsets might be most effective facing west, with the darker bedrooms positioned against the building's northern walls.

As discussed previously, encouraging solar radiation into our homes is only the first step. Capturing it in thermal mass is the masterstroke. Be it a concrete countertop, an earthen floor or a stone wall, the stored heat will constantly give back to your living spaces throughout the day and long into the night, after the sun has gone down.

Innovation in insulation and heating systems has made it possible to build homes without having to consider solar orientation. We have been able to rely on other fuel sources to provide all our heating needs. While it is possible to keep buildings warm using these other means, we are surely not at a point where we can discard the free, carbon-neutral energy that is provided by the sun. Just because we have learned to drive does not mean we should forget how to walk. Surely we should be doing everything we can to reduce the ongoing energy consumption of our homes, while simultaneously keeping ourselves connected to nature and the movement of the sun.

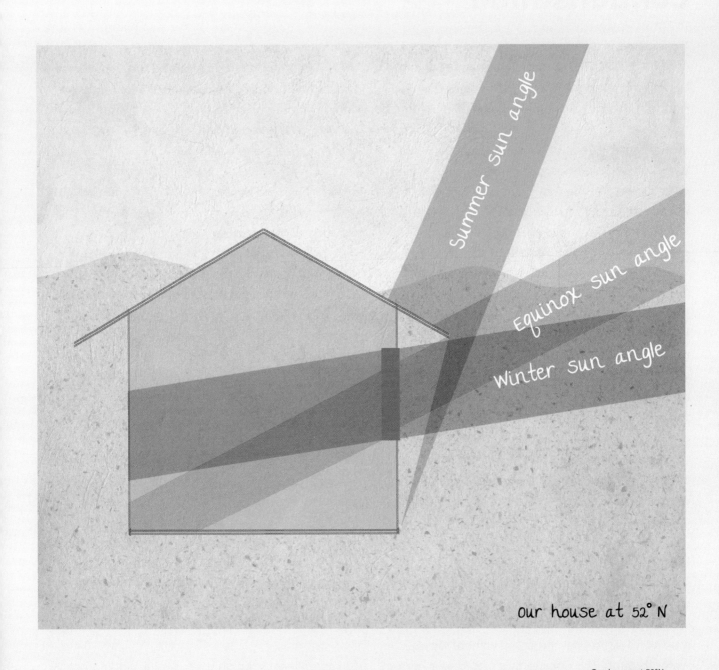

Summer sun angle

Equinox sun angle

Winter sun angle

our house at 52° N

Our house at 52°N, interacting with the various angles of the sun.

Understanding condensation

Condensation is the process of water vapour changing back from a gas to a liquid. It is the opposite of evaporation. Most air contains water, and we are very aware of this concept in Ireland, where we have an average humidity of 83% throughout the year. Condensation occurs when warm air encounters a surface temperature below what's known as the dew point, and the water vapour within the air collects on these surfaces. You will most likely have seen this phenomenon at play on the outside of a cold glass of water in a warm room. In a similar way, our windows are often fogged up with condensation in the morning, when the air temperature inside the house is warmer than the air temperature outside. All the moisture in the house is drawn to the cool surfaces. When it occurs within the walls of your house, it is called interstitial condensation. We went through a brief period in the past of lining the internal structural walls of our old stone cottages with plastic to prevent the damp condensation coming in. Unfortunately, this also prevented the moisture created within the house escaping, and led to problematic cases of mould and mildew.

In modern construction we have a number of solutions to prevent these issues, the most common of which is to use a breathable membrane. This is essentially a paper-like material that allows gases and air to pass through it without letting water through. This ensures that our spaces remain dry. Usually placed underneath the cladding on the external surface of a building, it prevents the humidity of the outside air penetrating the walls and potentially damaging the materials inside. It has become common practice to also line the internal walls with this barrier, to ensure that the moisture created inside the structure – whether from cooking, bathing or even breathing – does not soak into the building's walls and insulation.

Different materials deal with condensation in different ways, and some require less reliance on additional materials, such as breathable membranes. Earth plaster, lime plaster and natural stone walls actually absorb condensation and evaporate it back into the air and so do not need breathable membranes – they are able to regulate internal moisture movement themselves. In the last century we have developed an extensive array of new materials, none of which has the same naturally occurring breathable qualities of these traditional materials.

Modern construction has seen us replace the use of lime plaster with gypsum plasterboard inside our buildings. Plasterboard acts like a sponge; it is very good at absorbing moisture and does not readily release it, which can cause plasterboard walls to bulge or sag. In a country with such high humidity levels, it seems counterintuitive to line our walls and ceilings with a material that can so quickly cause mould growth, if not used or maintained correctly.

The short-term solution for this is the addition of extractor fans and the use of waterproof paint. Unfortunately, such paints often compound the issue of walls with zero-breathability and associated condensation. Lime-based paints are currently making their way back into popular use – with an impressive price tag – as they now come with all the necessary additives to make them behave like conventional paints in appearance and application. Of course, if gypsum is the best or only option you have access to, work with it, but understand its limitations and consider tile or lime plaster for areas that experience high levels of moisture and humidity.

Again, preventive measures are generally preferable to finding solutions, and it pays to keep this in mind when we are bringing large amounts of moisture into our living spaces. Steamy showers and the moisture they generate tend to attach to surfaces, so simply opening a window to the outdoors and keeping the bathroom door open will help a lot. It may seem obvious, but these simple ways of living in conjunction with nature get lost when we start relying on sensors and technology to alert us to excess moisture.

Condensation forming
on a single-glazed
window.

RECIPES

INSTALLING INSULATION

There is a full spectrum of insulation types available on the market – too many for me to list them all here. Instead it is my hope that by understanding how one insulation type is installed, you should be able to approach your preferred method armed with an understanding of the end goal.

All insulation types are designed to slow down the transfer of temperature, whether it is the cold from the outside or the warmth from the inside of our spaces. Some insulation is designed to lie flat, like a blanket, and others are designed to be more rigid, to allow them to stand up in a wall cavity. Insulation comes in different thicknesses, so be sure you are using the right thickness for your total wall depth. Do your research, and seek advice if you are unsure about what the correct type of insulation is for your project. And remember, the highest priority is slowing down the transfer of heat, so the more time you take to ensure that you have filled every gap, the greater success you will have.

Ingredients

- Suitable insulation. In this case we are using mineral wool insulation
- Knife
- Tape measure
- Handsaw

Safety note: Most insulation types are made up of tiny fibres, which can be extremely irritating to the skin, eyes and lungs. Always wear a dust mask, gloves and eye protection when working with these types of insulation. It is also recommended to wear long-sleeved clothes to protect your skin from irritation.

Method

1 Measure the distance between the studs that you are planning to fill with insulation. If your studs are placed at 400mm centres, and you used 50mm-thick lumber to build your wall, the gap between the studs will be 350mm. Add an extra 50mm to your measurement so that your insulation is always slightly wider than the gap. This ensures that it fits snugly between your studs.

2 Using the handsaw, cut a 400mm slice from your roll of insulation. Alternatively, you can unroll the insulation first and cut the insulation with a pocket knife.
Now measure the height of your cavity. Unroll your insulation and cut the strip to this length.

3 Install the insulation by holding it up and gently pressing it into the cavity, ensuring that it reaches both the top and the bottom. It should fit snugly, with no gaps. If you miscalculate your measurements at any point, and have any gaps, you can cut off small pieces of insulation and push it into the gaps.

MAKING YOUR OWN SHEEP WOOL INSULATION

Sheep wool used to be one of our most prized possessions, and sheep were originally farmed for their wool, not their meat. In fact, their meat was considered to be only fit for peasants. Today in Ireland, sheep wool is selling for such a low price that it is cheaper for our neighbouring farmers to give it away for free (or discard it), rather than to pay for the fuel required to drive it to the market!

The piles of wool that lie composting around the country are a symptom of a much greater problem on a much larger scale than the immediate need to insulate your own home. For now, let us focus on our basic human need for warmth and shelter, so that, from the comfort of our homes, we can address some of the bigger issues of the world.

We add boric acid to our wool to deter moths and other insects from nesting in it.

Ingredients

- Raw sheep's wool – usually available for free or at a very low cost, directly from sheep farmers
- Scissors
- Thermometer
- 3 large vessels – bins, buckets or basins with a capacity of at least 50 litres
- Eco-detergent and soda crystals
- Drying rack – you can make your own by spreading a wire mesh over a pallet and attaching it
- Boric acid solution

Safety note: It is necessary to wear gloves, a mask and eye protection when you are handling boric acid, and continue wearing them when installing your insulation.

Method

1 Spread your raw fleece across your drying rack and use a pair of scissors to cut out any particularly dirty pieces of wool. These off-cuts can go in your compost.

2 Lanolin, the oil present in the wool, is a fat and causes the distinctive, not unpleasant smell we associate with sheep's wool. It has a melting point of 65°C. (This temperature is just a little above what most hot water boilers produce for our showers.) Fill up your basin with hot water, at least 65°C, and submerge the wool. Resist agitating the wool too much, as this can make it start felting to itself. Just gently prod it and re-submerge it gently with your fingers, for about 10 minutes. Remove the wool from the liquid and gently wring it out before placing it in your second tub.

3 Wash your wool again in the second tub, which can be filled with either hot or warm water. This time you will be adding a capful of eco-detergent for every ten litres of liquid to your tub, as well as one capful of soda crystals. Submerge your wool again, gently prodding it, and let the detergent break down any remaining fats in the wool. Remove the wool and gently wring it out.

4 Prepare your third tub of water, using the same amount of detergent, and add your wool. Once it is suitably clean, lift the wool and rinse it in cool water. Now you can wring it out for the last time and spread it out evenly on your drying rack. You have now created a production line, and you can wash quite a lot of wool in the same series of tubs. Use your third tub as a guide to indicate when the water is too dirty and needs to be refreshed.

5 While the wool dries (and this may take a few days, depending on the air temperature), you can make a boric acid solution. Using protective gear, dissolve 1 cup of boric acid powder into 1 litre of boiling water and add it to a spray bottle.

6 Once the wool is dry, spray the boric acid solution all over the fibres. Turn the fleece to give it a thorough coating. Leave the wool to dry. Once dry, your wool is ready for installation into your building.

CHAPTER 5

WATER: FROM THE SKY TO THE GROUND

left Terracotta tiles in the Mediterranean.

Growing up in a drought-stricken Australia in the 1990s meant that I learned very early on that water is precious. The sound of rain on a metal roof still brings feelings of joy, relief and hope. The difference between prosperity and poverty came, not from the heavens, but from the clouds, and the scarcity of water was ingrained in our daily existence. Waterproof egg timers with small suction caps were sent out to every home in the country to help us limit our showers to three minutes, and conserve our most precious resource. Using outdoor taps was banned, and we were encouraged to report our neighbours if we spotted them watering their garden or – heaven forbid – washing their car, sending our glorious, life-sustaining, very limited resource down the drains.

I felt this same burden during the years I spent in India and Kenya, as monsoon season marked the survival of yet another year. I'll never forget the 42°C nights in Ahmedabad, sweating through the sheets, unable to sleep because of the heat and the noise of the ceiling fan clattering as it dangled off centre, threatening to break free from its chain at any moment. Another night-time memory joins this one, when the sky opened up and poured torrential rain down onto the city, flooding footpaths and ground-floor apartments, and drawing hundreds of pyjama-clad residents out into the street, dancing and yelping at the joy of rain on our skin. The rain lasted two days that season, and we wouldn't feel it again until the following year.

After so many years of travelling, I often wonder if I settled on the west coast of Ireland, not in spite of the weather, but because of it. The reliable, often seen as relentless, rolling in of rain clouds feels like an abundance I had not previously known. It seems obvious to me that in an ever-changing world, rainwater catchment in Ireland would be a simple and cost-effective method of ensuring water security. Our municipal water must be treated with chemical additives to allow for safe storage and transportation around the country, but such methods are not necessary in a domestic application, where we can store smaller quantities of water for less time, and use mechanical means of filtration, as opposed to chemical.

Moving water

While having an abundant supply of water for drinking, bathing and watering plants is certainly a luxury, ensuring that the areas we need to stay dry are kept dry requires careful planning and detailed work. We have come a long way from the leaking thatch and flagstone roofs of the past, and the development of various new roofing materials has allowed for much more comfortable living conditions. Let us explore the way that water moves, and the ways we can simplify our building designs to reduce the risk of water finding its way into unwelcome places.

Water, like most things in nature, will always take the path of least resistance. We can manage this characteristic by creating a sloping path for water to travel across our roofs and buildings toward the edges, ideally extending past the exterior walls, to move the water quickly away from our structure. This roofing extension is known as the eave. The further the eaves extend past the walls, the better, as it prevents wind blowing any falling water against our buildings. However, the further our eaves extend, the more sunlight is blocked out, another essential and precious resource which we want to harness. So finding the perfect balance of protection and acceptance of the elements is essential. Each and every house design is different, and the height of your windows greatly affects the interaction between the eaves and the sunlight. This is why it is so important to understand where the sun is going to be in relation to your home.

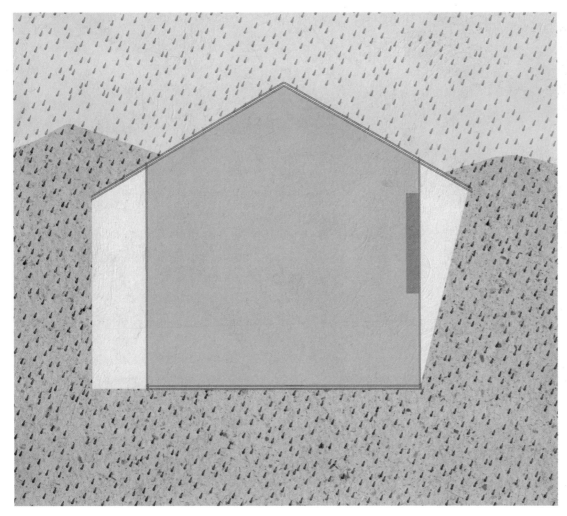

opposite Elegant floating gutter system in Connemara, Ireland.

left The effect of eaves in rain.

A technique called shingling is used to ensure that the path of water is continually moving down our roofs and away from our building. Shingling applies to all tile types, whether they are made from wood, clay, cement or stone. The basic idea here is to lay the tiles at the lowest point of your roof first, and then overlap the next layer of tiles slightly over the first. This process is repeated for each layer, until you reach the top ridge of the roof. If you are constructing your roof with corrugated steel, then shingling is not necessary, as you will likely be able to order the right length sheets for your roof.

When it rains, water is directed down along your sloping roof. Once the water has reached the last tile of your roof it should fall off into a gutter, or onto the ground. Sometimes, however, it will make a turn around the bottom of the last tile and begin climbing back up the underside of your roof, through a process called capillary action.

Often referred to as rising damp, capillary action is the ability of water to move, against gravity, through narrow spaces. It is the same force that allows plants to draw water up their stems from

surface. The more thermally conductive a material is, the faster it will change temperature and the more condensation it will attract. Steel roofing is a good example of this, although all roofing materials are made of conductive materials and will all cause condensation, to different degrees. We deal with this issue by adding a waterproof membrane beneath the outer roof surface, which collects the condensation and allows it to run off as a liquid. Our outer roofing material is then more a protective shell than a waterproof encasement. Advances in engineered fabrics have allowed us to replace the bitumen-infused papers of the past with a readily available material known as breathable membrane. It outperforms its predecessor on all fronts – longevity, strength, weight and price. We use wooden battens to create an air cavity underneath our external roof material to allow airflow to evaporate the liquid which gathers on the membrane. Building two layers of battens, in alternate directions, allows for further airflow and the removal of condensation from our roof.

the ground, and paper towels to wick up spilt water from a kitchen surface. To counteract this effect on our window and door sills, we add a drip edge. The drip edge encourages the water to group together, thus making each droplet too heavy to counteract the effects of gravity. It can either have a concave groove design or an additional edge piece; both succeed in making it harder for water to travel where we do not want it to go.

Condensation, yet another of nature's phenomena, can cause our valiant waterproofing efforts to feel somewhat futile. If you'll remember, condensation occurs when warm, moist air encounters a surface temperature below the dew point, causing the water vapour to condense and gather on the

Once we have encouraged all this water away from our structure, it is good practice to collect it in a guttering system and channel it all to one place, so that you can either harvest it for drinking or gardening or send it to a drainage system away from the house. If you do decide to collect and reuse your water, you will still need an overflow from your collection vessel so that any surplus water has somewhere to go. Drainage systems vary and range from entirely enclosed piping systems, which carry the water to a shared or municipal drainage system, to open landscaping, which channels your groundwater where you need it most. In either case, the most important thing is that the water is being taken away from your structure.

Living roofs

Living roofs are vegetated roof covers which can replace ordinary roofing materials. They add a protective layer of foliage to the structure, which increases its thermal resistance while protecting the building from wind, rain and UV damage. Living roofs also act as a filter for the rainwater they harvest and provide a home and habitat for local flora and fauna. The obvious reason why many people are attracted to living roofs is that they help our homes blend beautifully into our surroundings and exist in harmony with nature. There are two main types of living roofs: intensive and extensive.

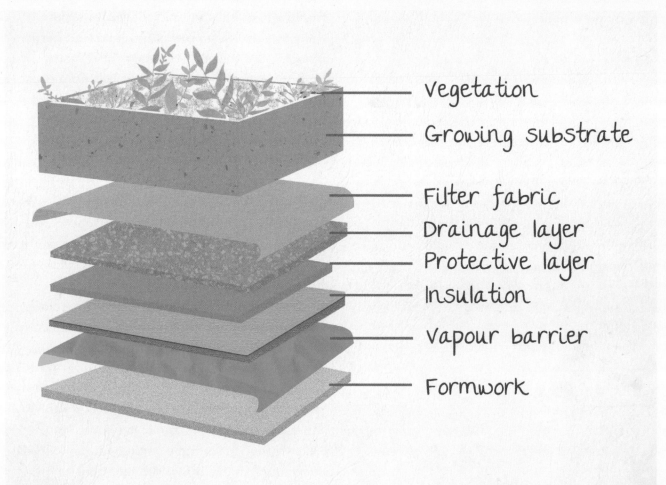

Vegetation
Growing substrate
Filter fabric
Drainage layer
Protective layer
Insulation
Vapour barrier
Formwork

above A living roof.

opposite Dealing with large amounts of rain in Connemara, Ireland.

INTENSIVE

Intensive living roofs are built over an EPDM (a type of rubber) lining. They can be installed on flat roofs and angled roofs up to a 30° slope. They very closely imitate nature, being made up of substrate layers of drainage gravel and 200–1000mm of soil. They can be planted with grasses, shrubs, berries and even trees. They can be extremely heavy, especially in wet climates, so a suitable roof structure must be built to allow for the weight.

EXTENSIVE

Extensive living roofs are also built over an EPDM rubber roofing lining, but can also be retrofitted to existing roof structures, such as corrugated metal roofing. A root barrier and geo-textile layer is laid first, which provides a drainage system. This is followed by a predominantly inorganic growing medium. Extensive living roofs are commonly planted with mosses, light grasses and drought-tolerant plant varieties, as there is no soil to store water in. These roofs are lightweight and usually require no additions to a roof structure.

The French drain

It can be useful to understand that water is, essentially, lazy. While capable of capillary action, it will always choose to travel via the path of least resistance. The key to ensuring that water does not move upward through your structure is to provide it with a more convenient path to take. Damp proof coursing is currently used in brick and cement block buildings and damp proof membranes beneath our floors are the current methods for resisting rising moisture from the earth into our homes. Rather than relying solely on such plastic and lead barriers beneath our structures, it is wise to utilise some more low-tech methods, which work in harmony with the physics of water, rather than simply trying to block its path. If you aren't considering where the water will go after you have prevented it entering your building, you will likely regret it!

Creating an underground drainage system known as a French or footing drain allows us to work with nature and the natural path of water. These drains can be essential in maintaining dry foundations and, therefore, dry walls and homes. French drains are a valuable ancient and low-tech drainage solution, traditionally made by digging a downward-sloping ditch, which was filled with coarse drainage rock. The slope was dug in the direction you wanted the water to travel and the drainage rock would allow the groundwater to move through it easily, and travel along the bottom of the drain. The ditch was dug to the same depth as the building's foundations, so that any water that could be tempted to travel up through the foundations would instead take an easier path, travelling sideways into the drainage system and away from the structure.

opposite Giving the water a path to the ground.

Over the years, a few simple innovations have improved the effectiveness and longevity of this drain type. Depending on your soil content, you may choose to line the bottom of your drain with a plastic membrane, to encourage good water flow away from the building. This is not necessary if your soil has even a moderate amount of clay content, as the clay will provide a smooth surface for the water to travel along. Just ensure you are precise with the installation of membranes within your drainage system – you do not want to block the path of water. It is important to always consider where your water is coming from, and where you want it to go.

Another useful addition to our modern French drain is perforated drainage pipe. Holes in the sidewall of the pipe allow the water in and it can then flow quickly and easily away from our structure, as opposed to having to navigate around the gravel the whole way along the drain. Using the correct

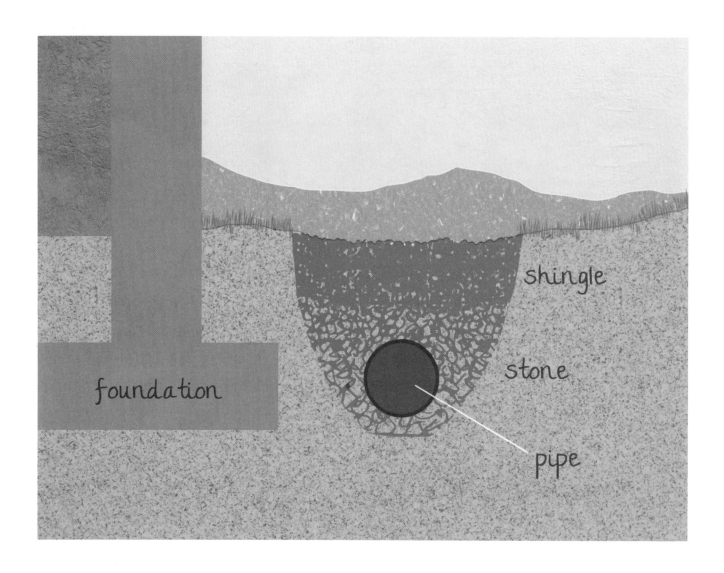

size of drainage rock will ensure good percolation. If your gravel is too fine the water will have a hard time getting through it, and small gravel particles can block the holes in your drainage pipe. If the gravel is too large, soil and debris can get between the rocks and block the path of the water. It is advisable to protect the top of your gravel with weed matting or cloth, which will allow water to pass through it while preventing soil and grass roots blocking your drainage system.

opposite French drain shown in section.

below French drain profile, showing 1/80 slope.

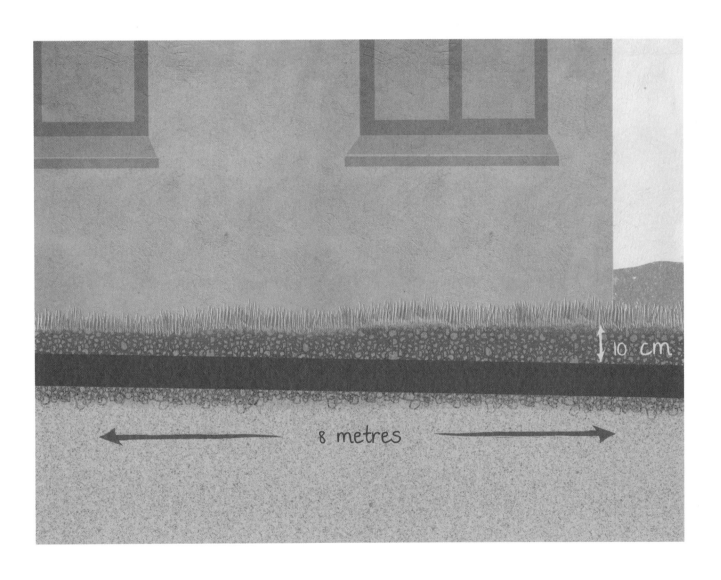

8 metres

10 cm

RECIPES

VAPOUR BARRIER

Vapour barrier – also known as a breathable membrane – is an engineered fabric, which has replaced the traditionally used tar-infused paper. It works as a layer of protection against any condensation or humidity that might make its way through the outer shell of your wall, roof and cladding material. The perforated holes in the vapour barrier are small enough to prevent water travelling through them, but big enough to allow air and gas to pass through. It is a highly effective protective layer, which is low-cost and easy to install. Vapour barriers are only used on the walls of a new build, as they are not appropriate for use over an existing masonry building.

Ingredients
- Roll of breathable membrane. Usually provides approximately 75m² of coverage
- Staple gun and staples
- 50 × 25mm pressure-treated timber battens
- 50mm × 4.0mm screws or nails.

Method

1 It is important to start your vapour barrier at the bottom of your structure so that each subsequent layer can overlap the previous layer and allow a clear path for the water down to the ground. Start in a place where there is a break, such as a doorway, as you will be able to continue this layer right around the building. Staple one end of your membrane to the timber frame, making sure you keep it straight.

2 Once it is securely attached, start unrolling the membrane around the building, in a similar fashion to wrapping a present. Use a staple to hold it in place as you unwind the roll. Ensure that the membrane is pulled tight, and is flat to the building as you wrap it. Now add a staple to the top and bottom of each stud, as you work your way along. Continue unrolling and attaching the vapour barrier sheet until you arrive back where you started. If you meet any window or door openings, you will be temporarily covering them with the vapour barrier.

3 When starting your next lap around the building, overlap the top of the first layer with the base of the second layer by at least 150mm, or as directed in the instructions of the membrane. This is to ensure that any moisture running down the membrane has a path right to the ground. Finish this layer like the first layer. Repeat the process until you have wrapped the entire building.

4 You can now add a vertical batten in line with every stud around the building. This will secure your membrane to the frame and will give a clear path for any moisture that may occur behind the cladding. These battens can be attached with screws or nails. Both will penetrate through your membrane, but the vertical battens ensure that the path for the water does not come anywhere near your fixings.

5 You can now cut into the membrane which is covering the opening of the doors and windows. I like to cut a Y shape from the top and an inverted Y shape from the bottom. This technique will leave you with two large pieces of membrane on either side of the opening, which can be wrapped over and stapled to the frame.

CHAPTER 6
FINISHES: CELEBRATING THE PROCESS

left Beautiful raw finishes in Daylesford, Australia.

above My father teaching me how to build in 1993.

I have always had an affinity for buildings with exposed structural elements. I feel that they celebrate the process which brought them to life.

My appreciation for the building process likely stems from my childhood, when I spent my days playing with the leftover construction materials which constantly sat in our yard as my parents built our family home. Neither of them were builders, but both had an interest in how things worked and happily spent ten years building the steel frame home that overlooked the white and blue ocean of Australia's Central Coast. I grew up running around on plywood floors and playing in piles of sand and gravel. Framed walls at their perfect 400mm spacing were everywhere and year after year, as the money was earned, another wall would be closed. I had grown up using a bathroom with exposed plumbing and open frame walls, and it wasn't until we sold the house that I ended up finishing that last bathroom, tiling it and installing a bath, before handing over the keys to the new owner. I have come to believe that my somewhat inherent knowledge of the construction process was passively absorbed during my years spent happily playing on that construction site, in a similar way as how the sun now passively heats my kitchen. I understand the length floor joists need to be, because I would jump across them as a child. I know how far apart wall studs should be spaced, because I have been squeezing through them my whole life. I had a good childhood, and my parents enjoyed that time as well.

My daughter's education in construction has already begun within this first year of her life. She has seen wardrobes being built in her room, and brick arches come together as she sits on a sheepskin outside, amused by her parents climbing up and down a ladder. She's seen steel beams and concrete countertops, and watched piles upon piles of sheep's wool being processed. She is passively learning a skill that is integral to her survival as a human. She is learning how to build shelter.

I believe that hiding so many structural aspects behind perfectly smooth walls is one of the fundamental causes of our generation's general confusion about how our structures are built, and how they work. There are so many more decisions to be made on the finishing layers of a building than just choosing paint colours. Many people are surprised to discover that they are not obliged to have plasterboard walls and ceilings, even if they have been strongly encouraged in that direction along the way. Imagine a home that has no secrets, where the bones of the structure are on display and the systems serving the structure are easy to follow, alter and repair when necessary.

With my personal building work, I feel motivated to create spaces that both a five-year-old and a 50-year-old can look at and understand how they were built, in the hope that this would help encourage their ability to construct their own spaces. Celebrating the way a building has been constructed, rather than hiding it, allows for a greater passive learning potential, and also provides us with an opportunity to be proud of and grateful for our design decisions and the qualities of our construction materials.

above Living with exposed structural elements subliminally teaches us how buildings work.

opposite Inari learning how the building comes together.

Exercise
X-RAY VISION

Look at the building around you and start to undress its many layers. Unless you were involved with the construction of this space, it may be hard to guess what exists behind the walls, if they are dry-lined with plasterboard. Can you guess what paths the power sockets or water supply have taken to get there? What elements are on display and where do they come from? How is your ceiling supported? Which walls are structural, and which are just acting as dividers? Could you potentially add a doorway through that wall, or remove the wall entirely to connect and enlarge your spaces? Such questions are exciting, intriguing and endless, and the answers so often lie behind 12 millimetres of flat, smooth plasterboard.

I am certainly no pioneer with such exposed design approaches. Structural beams, raw building materials and large open-plan spaces are all characteristics of the industrial architecture we have been familiar with since the 1700s. Without electricity, these historic buildings were tall and narrow with long windows, to allow for as much natural light as possible. Such tall buildings filled with heavily worked industrial machines posed a serious fire hazard, and the insurance companies at the time enforced strict building design regulations. These included the removal of any unnecessary ornate features, fabric curtains and carpets, the eradication of attic spaces and additional walls, and the end of flammable, oil-based wall paints.

The introduction of diesel-powered engines and the invention of reinforced concrete allowed for the expansion of such buildings. Architect Albert Kahn is famous for designing huge warehouses that could fit entire manufacturing processes into one space, and these buildings informed the design of our modern-day industrial buildings. Many of these warehouses were later converted into the studio apartments and workspaces that are so desirable today. Their open-plan design and the exposed structural elements originally chosen for their affordability and safety allow such spaces to be easily repurposed with exciting possibility and potential.

Usonia 1 – A naked building

At the height of the Great Depression, one of America's most famous architects, Frank Lloyd Wright, was commissioned to build a modest home which was simple to construct and affordable for most people. The resulting house would become known as Usonia 1. Designed to allow for maximum efficiency at low cost, it displayed simplified construction techniques and passive house approaches. Wright's unconventional approach saw him eliminate the use of plasterboard within the house, something very new at the time. He also reduced the amount of material used by creating an open-plan layout, maintaining that the only way to create such an affordable home was to allow the structural materials to simultaneously exist as the building's finishes.

With Usonia 1, Wright introduced the use of exposed internal brickwork, wood panelling, domestic track lights and underfloor heating – none of which had been seen in domestic dwellings before. In his efforts to cut costs and create a humble, democratic form of housing that could be emulated across the country, Wright broke from the norm and introduced the concept of celebrating exposed, natural materials, and this would influence American architecture for the rest of the century.

above Exterior
of Usonia 1.

over Interior of
Usonia 1.

Wall and floor finishes

We interact most with the final layer that our structure is wrapped in. It is the texture we feel when we lean against the walls and the surfaces we will spend our days sweeping and cleaning. Indeed, it is what our friends and family will visualise when they think of our home. When it comes to making decisions on which finishes suit us best, our priorities will differ. Aside from the consideration of cost, we must think too of the installation process, the environmental impact and the level of maintenance. Above all, external surfaces need to be hard-wearing and weather-resistant. The final shell is what protects the substructure of your home from the elements – rain, wind and animals. It is of equal importance to ensure that the external surface of your house is functional as well as beautiful, although if you find yourself having to choose one over the other, choose function!

opposite Gorgeous raw cladding in the Longhouse, Daylesford, Australia.

CEMENT RENDER

This has become a forerunner in Ireland as the conventional builder's choice for external plaster finishes. Its lifespan and minimal upkeep make it a popular option, although its ecological footprint raises questions about its suitability in what we hope to be a greener future. Cement is technically porous, but it slows down the passage of water to a great-enough extent to protect a structure from rain. Unfortunately, the same characteristic allows it to trap moisture within a wall, if it enters by some other means. Despite its strength, cement plaster is still prone to cracking over time. This is due to temperature changes and the expansion and contraction that naturally occur as it absorbs and releases water. Cement is made by heating limestone, sand and clay to temperatures around 1450°C in a rotating kiln. What is produced is known as clinker, which is then ground into cement powder. This process permanently alters the state of the material, and when hydrated the powder will become solid. Cement makes up only a small portion of concrete, accounting for 7–10% of the final mix, the rest being made up of sand, gravel and water.

LIME PLASTER

Lime plaster is a more ecologically sensitive plaster option, suitable for indoor and outdoor use. It is also made by firing limestone, but to a slightly lower temperature of around 1100°C. When heated, the limestone releases CO_2, becoming hydraulic lime or quicklime. Adding water to pure quicklime will turn it into hydrated lime (which can also be bought in bags from a builders' supplier, although bagged

the lime cycle

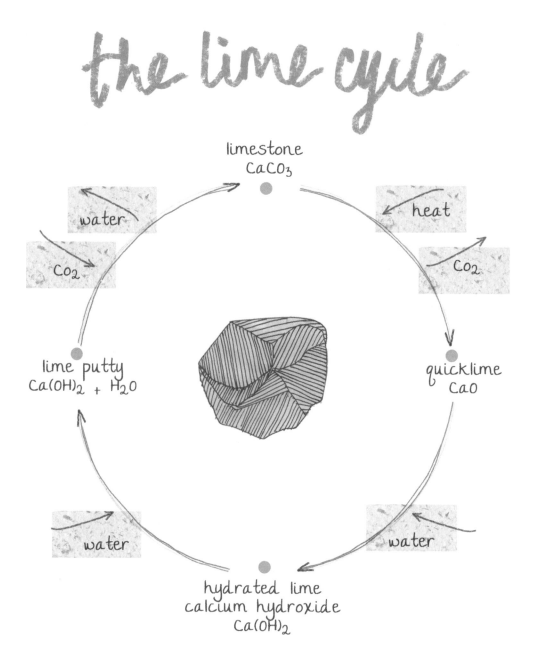

limestone
$CaCO_3$

water

CO_2

heat

CO_2

lime putty
$Ca(OH)_2 + H_2O$

quicklime
CaO

water

water

hydrated lime
calcium hydroxide
$Ca(OH)_2$

hydrated lime usually also includes other additives). However, mixing pure quicklime and water results in a vigorous reaction and gives off heat, so should only be carried out following appropriate safety guidelines. As lime mortar dries, it absorbs carbon from the atmosphere and turns back into limestone, in its new shape. As the chemical bonds within the lime are not destroyed during the heating process, we can reconstitute it with water to turn it back into a usable lime product. This quality is one of the advantages of lime over cement, which can never be brought back to its original elements.

GYPSUM

A naturally occurring material which is relatively easy to mine, process and install as an internal wall covering, gypsum is processed or 'cooked' to create builders' gypsum for plastering, which is set into flat sheets held together by a paper outer layer, to become what we know as plasterboard or dry lining. Plasterboard became popular in the mid-20th century and, due to its low cost and ease of installation, it has become ubiquitous in the construction industry. Its disadvantages are that it is very fragile and susceptible to water damage. Since its surge in popularity in the 1950s, there have been countless interventions to improve the qualities of gypsum board. Efforts have been taken to make it soundproof, water resistant and fire retardant, with varying rates of success, and it has provided the perfect background for the incredibly lucrative paint industry.

TIMBER CLADDING

This has been a popular option for sheathing the exterior and interior walls of homes for hundreds of years, and it is still in use in many parts of the world where wood is in abundant supply. It is insulative and, while it is not waterproof, if installed correctly it has the ability to absorb and release moisture without being damaged – the key thing here is that the timber must be able to release its moisture. It is most easily applied to internal timber frame walls and can be painted, stained or left raw. Attaching wood cladding to your stud walls with visually appealing rounded nails can make a feature of the process, while also helping you to know where the timber studs are positioned.

There are many suitable cladding timber varieties to choose from, with varying qualities. Cedar is naturally mould-resistant, while pine and larch require treatment to prevent their open grain structure acting as a surface for mould growth. Most external timbers will naturally turn silver or grey with age and UV exposure, and all can be painted or stained. My preferred method for external timber treatment is the Japanese technique of Yakisugi, which involves charring the wood and creating a charcoal layer which bugs, insects and mould cannot live on.

CLAY BRICKS

Clay bricks are a long-lasting interior and exterior sheathing option which are wonderfully durable and can withstand extreme weather conditions. Clay is one of the oldest and most plentiful building materials on the planet, and we have used and adapted it for many generations, constantly finding new uses for this versatile and abundant natural material. Clay bricks are made of fired clay, with a few additives to ensure the consistency of their texture and colour. Usually bonded together with cement or lime mortar, bricks are a great example of an easy-to-apply structural material which can also serve as a finished surface.

EARTHEN PLASTER

This is traditionally made of a combination of clay, fine aggregate, fibre and pigment. It offers an internal and external finish, although in an external setting it will require annual upkeep. It is able to absorb and, more important, to release moisture without disintegrating. It is a wonderfully forgiving and fun material to work with, and is therefore a great option for the novice plasterer. Clay naturally gives us warm, earthy tones, which can be altered with the addition of iron-based colour pigments. It should be oiled or waxed when used in wetrooms.

TILE

Ceramic tiles date back as far as 4000BC, with the earliest known sources found in Egypt. Since then, they have found their way into the hearts and homes of many cultures across the globe, being used both internally and externally on floors, ceilings and walls. The abundance of clay and the evolution of glazing techniques have allowed tiles to retain their popularity, and their dual functional and aesthetic service to us over the millennia. The tiles you might source now could be clay or cement-based, sealed or unsealed, printed, hand-painted or colour-glazed ... and this is only the beginning! When choosing tiles, it can be useful to first determine where they will be used and what surface you will be adhering them to. Floor tiles are often thicker than wall tiles; do you need them to be wipeable or resist heat? How many will you need? All of these factors will help determine which tiles are right for your project.

opposite, top Rain shedding from exterior cladding treated with the Yakisugi process.

opposite, bottom Stucco render in Santa Fe, New Mexico.

above Matte glazed tiles meeting a timber floor.

TADELAKT

Tadelakt is a variation of lime plaster which is ideal for wet areas, both internally and externally. Developed in Morocco for the steaming bath houses, tadelakt uses hydraulic lime, which is blended with marble dust instead of sand. As marble dust particles are significantly smaller than grains of sand, this plaster must be applied in very thin coats, which are then compressed and polished by hand. This achieves a beautifully smooth finish which should be regularly coated in olive oil soap (also called black soap) to make it water-repellent. Tadelakt is a labour-intensive process, but the results are worth the work.

Flooring considerations

Of all of the finishes in our homes, we come into more contact with our floors than with any other surface. Choosing the right floors for the right room can determine how you interact with that space. Keep in mind the practical aspects of cleaning and maintenance as well as creating a comfortable and attractive space.

My personal preference leans towards wooden flooring in the rooms where we are often barefoot, such as bedrooms and bathrooms, as it is warm to the touch and easy to keep clean. The rooms that experience heavy traffic, such as kitchens, porches and utility spaces, are often best laid with hard-wearing materials such as concrete – which has the added benefit of being thermally massive – tile or hardwoods, which you won't mind scarring and scratching. If you love being barefoot, wood, carpets and cork are the direction you should be exploring. If ease of cleaning is highest on your priority list, tiles and melamine will make you happy. We are all different, and our choice of building materials reflects this wonderfully.

left Exposed structural elements in an earthen building, Taos, New Mexico.

opposite Oxidised iron stain on a concrete floor.

RECIPES

LIME PLASTER

I have yet to find a disadvantage to hydraulic lime plaster. It meets all our requirements of weatherproofing and airtightness, while having none of the associated problems similar alternatives seem to create. Suitable for both interior and exterior use, it is breathable, firm yet flexible, less likely to crack than cement (if applied correctly and if the relevant aftercare is carried out) and less likely to dissolve than gypsum. The high pH of lime acts as a natural fungicide, it requires less energy to create than cement and can always be reconstituted back into a usable product. We have a lime-plastered bathroom and have never seen mould growth after five years of steaming the room out with hot showers. It is safe to say that I am a big fan of this product!

Ingredients
- Natural hydraulic lime 3.5
- Sand
- Water
- Cement mixer
- Sponge
- Trowel and hawk
- Wheelbarrow
- Scratcher
- Hessian and nails (if required)

Safety note: It is very important to wear protective equipment when working with lime, as it is a highly alkaline powder. Always wear thick rubber gloves, a suitable dust mask and eye protection. If you feel any irritation on your skin when using lime, rinse the affected area with vinegar, which will counteract the burning alkaline effect of the lime.

Method

1 When applying plaster to a raw masonry wall, it is good practice to first apply a very thin scud coat to the wall the night before plastering. This will help the plaster adhere to the wall and allow for even coverage. Make up a 1:1 lime-to-sand mixture and apply it roughly across the wall by either flicking the mixture from your trowel or spreading it thinly with your hands or your trowel.

2 The ratio for our hydraulic lime plaster is a 1:3 lime-to-sand mix. The water content is never a specified amount, but it will be approximately ¾ of one part. Add half of this amount of water to your spinning cement mixer. Add 1 part of your sand and then add your lime. This should create a wet mixture that is fully incorporated. Let this mix stand for a minute or two.

3 Add another part of your sand and observe the mixture. If it looks too granular, add a little more water, and always allow the water 30 seconds to integrate before adding more. Keep in mind you have one more part of sand to add, so a mix that looks a little too wet at this stage is a good thing.

4 Add the remaining sand, and let the materials combine. Add a little water as necessary, until the mix has a smooth consistency, similar to soft-serve ice cream. The mix is now ready, and can be poured into a wheelbarrow or large bucket. It should be used within the next few hours or it will carbonate/set.

5 Before we apply any plaster to our wall, it is very important to wet it down first, using a sponge, hose or spray bottle. The substrate should be damp throughout, but not dripping or glistening with water. There is a fine balance to this. If the substrate is too dry, it will absorb the moisture from the lime too quickly, and the lime will not cure properly. If the substrate is too wet, it will form a film between the lime and the substrate and prevent good bonding.

6 Use your trowel to drop a mound of plaster onto your hawk, then scoop up a small amount of plaster on the end of your trowel and apply it thinly to the wall. Aim for a plaster thickness of between 3mm and 7mm. It can be applied thicker than this, but it is more likely to crack as it dries, so it is better to apply multiple thin layers, rather than a single thick layer. Repeat the process, plastering in an upward, diagonal motion and smoothing the seams regularly.

7 After about an hour, gently scrape the surface of the plaster with the scratcher, creating grooves so that your second coat tomorrow will have a rough texture to adhere to.

8 Allow it to cure for 24 hours before applying the second coat.

9 Between one and two hours after you have applied the second coat of plaster, you can use a damp sponge to gently smooth over and float your wall. This process allows you to merge any seams and gives a soft yet textured finish, as this floating process brings sand to the surface.

10 If applying lime plaster in direct summer sun or in wet conditions, it should be protected by hanging sheets of damp hessian over it, tacked lightly at the top of your wall with nails, to prevent the lime drying too quickly and cracking. Avoid using lime if the temperature is below 5°C, as it will not cure/set properly and will crack.

EARTHEN PLASTER

Making earthen plaster is a wonderfully natural and instinctive activity, reminiscent of playing in muddy puddles and sandpits, as a child. Put on some old clothes and enjoy the game of finding the perfect balance between the stickiness of the clay and the structure of the sand, with the gorgeous flecks of golden straw throughout. I believe building should be fun, explorative and experimental, and earthen plaster is exactly that.

Ingredients
- Strained clay (see the cob recipe on page 131)
- Sand
- Bucket
- Straw, chopped up to approximately 1cm pieces
- Cement mixer
- Water
- Wheelbarrow
- Trowel and hawk
- Scratcher
- Hessian and nails (if required)

Method

1 The ratio I like to work with for clay plaster is a 1:2 clay-to-sand mix. Depending on how wet your strained clay is, you may not need to add any water at all. Add one bucket of sand and one bucket of clay to the mixer, let it spin for at least a minute, and assess your mixture. You want it to be quite wet at this point, similar to a milkshake consistency. Add a little water, if necessary. Now add the second bucket of sand and assess the mixture again. You are aiming for a final consistency that resembles soft-serve ice cream – stiff enough to hold its shape, yet soft enough to spread with a trowel.

2 Once you are happy with the consistency of the mix, add a large handful of chopped straw to your mixer and wait for it to fully merge with the clay plaster, so that the straw is evenly dispersed throughout the mixture.

3 Once all the straw is blended into the clay/sand mix, empty the plaster into a wheelbarrow.

4 As with lime plaster, the substrate you are applying your plaster to must be thoroughly damp before application. You can use a spray bottle, sponge or hose for this. It should be wet but not glistening to ensure the best adhesion possible between our plaster and our substrate.

5 Use a hawk and trowel, or your hand if you prefer, to spread the mixture over your wall, pressing firmly into any gaps between stones, bricks or whatever your substrate is. Aim for a 3–5mm thickness. If it is too thick, it will crack as it dries.

6 Systematically apply the plaster to the wall using your trowel. You want to create a smooth, even finish, so pay special attention to any seams or trowel marks, smoothing them as you go. Try to create an overall, even surface. Scrape the surface using a scratcher and leave your plaster to cure overnight.

7 Repeat the process on day two, again leaving it to cure overnight.

8 On day three, apply the earthen plaster, and instead of scraping the surface, wait for 1–2 hours for the wall to begin drying. Then you can use a damp sponge to gently smooth over and float your wall, which will give it a soft, textured finish. Alternatively, you can create a smooth surface by repeatedly trowelling the wall as it dries. It will harden as the water evaporates and as you continue to compress the plaster with your trowel.

9 If applying clay plaster in direct summer sun or in wet conditions, it should be protected by hanging sheets of damp hessian over it, tacked lightly at the top of your wall with nails, to prevent the clay drying too quickly and cracking. Avoid using clay if the temperature is below 5°C, as it may crack if the water inside it freezes.

LIME PAINT

Humans have been making paint for as long as we have existed. We use paints to express our emotions, our style and our creativity. We use it to cover up stains, and to make spaces appear clean and new. Unfortunately, most standard paints contain a multitude of harmful chemicals with the potential for ecological and toxicological damage during their production, application and especially their disposal. They contain hazardous levels of VOCs – volatile organic chemicals – which can lead to indoor air pollution and knock-on adverse health effects. The good news is that we do not have to do without the exciting possibilities of paints, stains and varnishes, as there are a multitude of natural solutions available for minimal cost.

Here is an extremely simple and low-cost recipe for making your own beautiful, breathable, anti-fungal paint. The addition of earth-based pigments is optional. A natural, unbleached lime usually provides a soft, oatmeal colour. If you are using a pigment to tint your paint, you will need to find one that is mineral-based so that the mineral lime will accept the pigment. Clay-based pigments work well, as do iron-based pigments. I tend to use what I can find locally, but there are a multitude of earth-based pigments available to source online or through natural paint suppliers.

Only make as much paint as you need in one day, as hydraulic lime reacts with water and then carbonates as part of its drying process.

*Note that lime wash and lime paint are similar but different, and their uses are not interchangeable. The following is a recipe for lime paint.

Ingredients

– 1kg of natural hydraulic lime (which can be bought in three strengths, 2, 3.5 or 5; any will work for this)
– 2 litres water
– 200g earth-based mineral pigment (optional)
– Drill with a whisk attachment
– Hose or spray
– Masonry brush

Safety note: As is always the case with any lime work, safety glasses, thick gloves and a protective dust mask should be worn. If you feel any irritation on your skin when using lime, rinse the affected area with vinegar, which will counteract the burning alkaline effect of the lime.

Method

1 For this recipe, I have used unfired brick clay to add a warm terracotta tone to the lime paint. If working with dry pigment, I recommend slaking or soaking the pigment in water the night before. This will help to loosen the pigment and make blending it easier. If you are making paint in a hurry, you can skip this process and use an electric blender to liquify the pigment.

2 Add your water to the bucket and then slowly add your lime powder, whisking as you go to incorporate the lime powder into the water. Once all the lime has been added, you should have something resembling double cream, or store-bought paint for that matter!

3 Next, add your pigment slowly, whisking continuously until your desired colour has been achieved. The paint will dry substantially lighter in colour than what you see when it is wet.

4 Wet down the surface you intend to paint using a hose with a mister nozzle, or a large spray bottle. This will ensure proper adhesion.

5 Brush your lime paint on thinly with a large masonry brush. If you apply the paint too thickly it will clump and crumble off when dry. Much like applying lime plaster, you want to apply several thin coats, as opposed to one thick coat. Stir the paint regularly.

6 It is recommended to leave at least 12 hours between coats to allow the lime to cure. If you intend to apply more than three coats to your wall, it is recommended to allow the wall to cure for 5–7 days after the third coat. This allows the paint to absorb the carbon dioxide from the air, and to harden. Once the reaction has finished you will effectively have a very thin, coloured layer of limestone on your walls.

TADELAKT

Tadelakt is a traditional Moroccan lime plastering technique which produces a beautiful polished, waterproof finish. It is a labour-intensive process, but the results are worth the effort. It is made of hydraulic lime, with marble dust as its aggregate. Because the marble dust is so fine, it must be applied in extremely thin layers to prevent cracking. It also means that it is possible to compress and polish the lime after it has cured, to create a wipeable, non-porous surface. You can either source authentic Moroccan tadelakt powder, or you can make your own powder using a hydraulic lime and marble dust mix with a ratio of 1 part lime to 4 parts marble dust. Black soap, also known as olive oil soap, is used to waterproof the plaster. It does this by penetrating into the plaster and blocking any open pores, which prevents water from entering.

Ingredients

– Moroccan tadelakt powder, or hydraulic lime and marble dust to make your own
– Cement mixer or drill with a whisk attachment
– Sponge or spray bottle
– Hawk and trowel
– Scratcher
– Smooth, round, palm-sized polishing stone
– Black soap, also known as olive oil soap

Safety note: As is always the case with any lime work, safety glasses, thick gloves and a protective dust mask should be worn. If you feel any irritation on your skin when using lime, rinse the affected area with vinegar, which will counteract the burning alkaline effect of the lime.

Method

1 Tadelakt, much like any other plaster, should be prepared using a cement mixer or a mixing tool. If you have authentic tadelakt powder, you only need to add water until it reaches the desired consistency. It should be thick but not holding any form, similar to custard.
2 Wet down your substrate before applying the tadelakt, using a sponge or a spray bottle with a fine mist nozzle. It should be damp but not glistening wet to ensure the best possible adhesion.
3 Starting at the bottom of your working area, apply the tadelakt to the substrate with a trowel. It should be applied as thinly as possible, no more than 2mm thick.
4 Continue to apply the plaster, working outwards from

your starting point. Always come back to blend the seams and trowel marks as you work.

5 Once you have covered the entire area, lightly scratch the surface of your tadelakt plaster to create a good surface for gripping onto with your second coat. Leave to cure overnight.

6 The following day, trowel another thin layer of tadelakt over your working area. Take care to push the plaster on firmly, and remove as many trowel lines and indentations as possible. Once you are satisfied that you have created the smoothest possible finish, leave the tadelakt to cure.

7 Check the plaster every hour by trowelling a small portion of the wall. If the plaster seems dry, trowel firmly over the same spot several times. This process is called hard trowelling, and it will bring moisture to the surface. Leave the area to fully cure for at least 24 hours.

8 After 24 hours, test a small area of the plaster with your polishing stone. You should be able to press and buff in small circles to create a shiny polished surface. If you notice the plaster crumbling or breaking away, it needs further drying time, so wait another 24 hours.

9 The polishing process is the true labour of love with tadelakt. It can take up to an hour per square metre of tadelakt to polish with the stone. You are essentially compressing the marble dust and lime and creating a non-porous surface. Keep your pressure and pattern consistent for the best results.

10 Once you have finished polishing the tadelakt you can dilute your olive soap with three parts water, and wipe it over the walls. The olive oil soap should be reapplied once or twice a year.

TILING

Tiles can instantly transform a space, as well as being waterproof, stain proof and extremely hard wearing. Ceramic tiles are the grandmother of pottery, and every culture had its own unique style and techniques for making tiles. Tiles are one of the areas of building where most of the work has been done for you – you just need to set them in place.

When choosing your tile adhesive, pay attention to the area you are planning on tiling, as different adhesives are designed for different surfaces and temperatures. Most tile adhesives have a working time of only around 30 minutes, so only mix as much as you think you will use.

Ingredients
– Tiles
– Tile adhesive
– Drill with mixer attachment
– Tiling trowel
– Tile spacers
– Large sponge and dry cloth
– Grout
– Rubber grout trowel
– Tile-cutting saw or an angle grinder with a diamond blade

Safety note: Always wear safety glasses, gloves and ear protection when using a tile cutter or angle grinder to cut tiles.

Method

1 Ensure the area you are planning to tile is both clean and stable. Tiles like a solid backing. For best adhesion your tiles should be soaked in a bucket of water before you use them. I like to soak ten tiles at a time and then have them ready nearby, with the excess water draining away. Mix your tile adhesive according to the directions on the packet.

2 Use your tiling trowel to plaster your tile adhesive to a small area of your wall. I suggest starting at the floor or in a corner so as to get your pattern right.

3 Use the teeth of your tiling trowel to create grooves in the tile adhesive. These grooves allow you room to push your tile to the correct depth, and in line with the neighbouring tiles.

4 Take your first tile from the soaking bucket and place it onto the tile adhesive. Press it gently, until you are happy with its position.

5 Place the next tile, and put in a tile spacer, to ensure you have an even and consistent space between the tiles. The gap is to allow for expansion and contraction of the tiles.

6 Continue placing your tiles and tile spacers and spreading your pattern outwards.

7 You may soon reach the limit of your tiling space, and will be required to cut a tile. Use an angle grinder with a diamond blade attached or a tile-cutting saw. Ensure that the tile is wet before you cut it, and work slowly.

8 Once all your tiles are in place, remove your spacers and gently wipe the surface to remove any excess tile adhesive. Take care not to shift your tiles out of position. Leave it to set overnight.

9 The next day, mix your grout as per the instructions on the packet.

10 You can now apply the grout with a flexible rubber grout trowel into all the gaps between the tiles. I like to grout a square metre of tiles at a time, wiping the tiles with a sponge and dry cloth as I go. Refer to the recommended drying time on the packet – usually overnight.

TIMBER CLADDING

Exterior timber cladding is an attractive and easy-to-install solution to protecting our buildings from the elements. Certain wood types are more suited to life outdoors – larch and cedar are popular, albeit expensive, options. There are ways of preserving more affordable wood types to make them more weather-resistant, such as Yakisugi wood charring – see page 236.

In the following method we will look at a vertical and a horizontal timber cladding installation to show the slightly different approaches they require. The horizontal cladding we are using is pressure-treated pine and I recommend using what is known as shiplap boards, which rest on top of each other and overlap.

Safety note: Always wear eye protection and gloves when using a saw.

Vertical installation – larch

Ingredients

– 225mm × 25mm suitable timber cladding – in this case, we are using Irish larch
– 75mm × 25mm cover strips, to match your cladding
– 90mm exterior-grade screws
– 40mm exterior-grade screws
– 50 × 25mm pressure-treated battens
– Saw
– Hammer or impact driver

Method

1 To install vertical cladding, you will need to add a layer of horizontal battens to the structure, to which you can then attach your cladding. For this method, we are using 50 × 25mm pressure-treated battens installed at 1-metre increments up the wall. It is essential to add vertical battens first, over your membrane, directly in line with the studs in your frame. These vertical battens ensure good airflow and drainage for any potential moisture that may occur behind your cladding. You can use external-grade 90mm nails to attach the horizontal batten at every point it intersects with a vertical batten. This will also ensure that it aligns perfectly with your wall studs.

2 Now we can install our 225mm × 25mm cladding side by side across the face of our structure. Ensure your first board is perfectly vertical, or plumb, and then attach it with two nails or screws on every horizontal

batten with the 40mm nails or screws. If you keep your nails or screws close to the edges they will be hidden in the finish look. Place the next board beside the last piece of cladding and repeat the attachment process. Continue your way around the building until all your cladding is installed, cutting any pieces to fit around doorways or the corners of your building.

3 Next, take the 75mm × 25mm cover strips and centre them along the seam, where each vertical piece of cladding timber meets the next piece. All wood expands and contracts with humidity and temperature fluctuations, so we compensate for any gaps with our cover strips. Use the 90mm fixings to attach one screw or nail through the cover strip to each horizontal batten, all the way to the top of the wall.

Horizontal installation – pressure-treated pine

Ingredients

– Pressure-treated pine shiplap cladding (typically 170mm in width)
– 40mm external-grade nails
– Saw
– Hammer or impact driver

Method

1 Position the first piece of cladding at the bottom of the wall and ensure it is in the right position and level. Use one fixing in the bottom half of the shiplap to fix the shiplap to the vertical batten.

2 Sit your next board on top of the last, making sure that the lap is resting over the top of the previous board and that the profile is lined up correctly. Use screws or nails in the bottom half of the board in every batten. Repeat this process until you reach the top of your wall.

3 If you are joining two pieces of horizontal cladding together on your wall, make sure that they join over a batten. Try to stagger these vertical joins, so that the pattern looks random and continuous across the wall.

YAKISUGI WOOD CHARRING

Yakisugi is the ancient Japanese technique of preserving wood by charring it, which has been practised for centuries. Our two main concerns when it comes to wood preservation are insect infestation and mould, or rot, as a result of moisture. Charring the wood deters insects that would normally be drawn to it, and at the same time carbonises all the live material on the outer layers of the wood, which makes it an equally inhospitable environment for any mould or fungal growth.

As well as its attractive appearance, the main advantage wood charring has over common copper-based wood preservatives is that it is a chemical-free process. This means that the wood can easily biodegrade back into the earth at the end of its use.

Ingredients

- Timber cladding
- Felting torch or similar
- Propane canister
- Wire brush
- Boiled linseed oil
- 10cm wide paintbrush

Safety note: Always wear heatproof gloves and footwear when using the felting torch. Pay attention to your surroundings, making sure there is nothing flammable nearby and that nothing catches fire.

Method

1 Prepare a safe, non-flammable space to work in. I like to lay my boards out on my gravel driveway. Lay the boards out a convenient distance apart so that you can walk between them.

2 Start torching your wood lightly by holding the torch above it and observing the flame. Slow, smooth passes back and forth at a regular distance from the timber should give you good results, much like painting. Adjust your distance from the wood and the speed of your torch to char the wood more, or less. To get the best preserving results from Yakisugi you must completely char the outside of the wood.

3 Once you have charred all your wood, brush the char very lightly with a wire brush to remove any loose charcoal.

4 You can now paint or wipe on your boiled linseed oil and allow it to soak into the wood.

5 Use the torch again to heat up the linseed oil on the wood. This will thin the oil and open the pores of the wood, allowing it to penetrate deeply. You will notice the slick oil sheen disappearing from the wood when you remove the torch.

NATURAL WOOD STAIN

My wife, Erin, is a textile designer, and she occasionally shares her natural dyeing secrets with me. I'd like to share my favourite with you here. Although altering the appearance of wood has never been a great love of mine, I believe there is beauty in accentuating the grain and texture of timber with natural stains. This recipe offers charcoal grey to light grey tones, depending on your wood type and the strength of your iron solution.

For a noticeable colour change to occur, the presence of tannin is required. Some wood types, such as oak, walnut, cherry and mahogany, are naturally high in tannins. In fact, the word 'tannin' comes from the Old German, word *tanna*, meaning oak. The recipe given here shows you how to add a tannin wash to any wood so that, regardless of its tannin levels, a colour change will occur. You can either use iron oxide powder for the stain or you can make your own iron solution by soaking steel wool in vinegar for three days before you plan to use the stain.

Ingredients

- Large jar
- Steel wool pads
- White vinegar (all types work, but maybe not balsamic, as it's sticky)
- 100g black tea
- Gloves
- Sieve

Method

1 To make the iron solution, fill a large jar with the steel wool pads and cover with vinegar. This should be left to steep for at least three days prior to using so that the iron starts to rust into the vinegar. The more steel wool you use, and the longer you leave it, the stronger your mixture will become. When ready to use, strain the steel wool through a cheesecloth or fine sieve. Be careful not to splash the iron solution onto anything which you don't want it to stain.

2 To create the tannin wash, boil a litre of water in a pot and add 100g strong black tea, either tea bags or loose tea. Allow the tea to steep for at least one hour before straining it. (You can skip this step if you are working with woods with a naturally high tannin content.)

3 Using a wide paintbrush, paint your tannin wash onto the wood, if using. Leave it for an hour or so, to allow it to soak and dry into the wood.

4 Next, paint on your iron solution. You should notice an immediate effect on the wood which will continue to change for up to an hour after application. Allow the wood to dry completely before lightly sanding, and sealing with wax or an oil.

INSTALLING SOLID WOOD FLOORING

Whether in a new build or a renovation, installing wood flooring is a relatively simple task. There is a trailer load of tools available to speed up the job, but the owner-builder on a budget can achieve great results with a drill, a hammer and a block of wood or plastic. Most flooring you will have access to now has a tongue and groove system to help create a seamless finish. The method below assumes you are applying floor to a suspended timber floor, i.e. a floor made of timber floor joists, already installed.

Ingredients

- Suitable flooring timber
- Drill and a 3mm drill bit
- Screws 3mm diameter × 40mm long, preferably with a small head
- Internal-grade wood glue
- Wood or hard plastic block

Method

1 It is standard practice to run the length of your boards in the long direction of your room or hallway, but they must be laid perpendicular to your floor joists below. Working onto your solid subfloor structure, lay out your first row of flooring along the edge or wall of your work area. It is important that the ends of the floorboards are always supported, so lay them so that all joins lie over a floor joist. If the boards you are using are thinner than 18mm or too short to always allow them to land on a joist, you can first add a sub-floor layer of plywood or OSB sheeting to give you a flat, even surface to work on, with potential attachment and joining points anywhere.

2 Lay the boards with the tongue facing in the direction you are going to be working. Once you are satisfied your first board is in the correct position, drill a screw through the timber flooring and into the joist below. Once you have your first row attached, line up the next row of floorboards. Squeeze a bead of glue into the groove of your next board.

3 Starting at one end, place the board in line with the tongue. You may need to use a hammer to gently knock the groove into the tongue of the previous board, and you should always use a block of wood or hard plastic between your hammer and the board, so that you do not damage your floorboards. Once the boards are in the correct position, drill and screw again.

4 Try to keep your joints random in the pattern they form, as they are easily noticeable on floors.

CHAPTER 7
THE ORDER OF
OPERATIONS

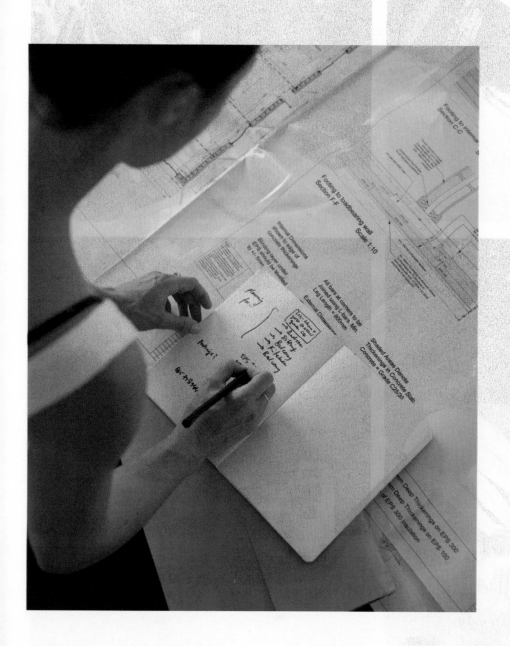

If you are an owner-builder it can be useful in the preparation stage of your build to understand the order of operations, or the sequence of events, before you dive in. But first let's take a look at the importance of reading and understanding plans and drawings, and reflect on the often disjointed approach taken to building our homes in order to fully appreciate the benefits of a deep engagement with the processes, people and material involved.

Reading plans

Construction plans and drawings are not difficult to understand; however, few people are given the opportunity to practise this skill. I want to encourage you to engage with your building plans before you assume that you won't be able to understand them. They are another useful communication tool on a construction project and another language to learn and use. Similar to geographical maps, construction drawings are a 2D representation of a proposed or existing structure, and they use a combination of drawings, references and icons to represent important design and structural information.

There are two types of plan you are likely to encounter during a construction project:

Architectural plans are designed to inform the client on the overall look of the building. These plans often include detailed elements, such as toilet locations, dishwasher placement, light fixtures, etc.

Engineer or construction drawings outline the relevant building methods, structural components and code compliance. Depending on the complexity of the project, the plans will be more or less detailed. So, to exemplify the difference between these two types of plans, we can say that an architect's drawings may show a wall and the construction drawings will show how the wall will be built, and with what materials.

A floor plan is an overview of your proposed structure. It will illustrate the overall footprint of the building, its orientation to the north, its relation to surrounding landmarks or buildings. It should also illustrate the thickness of your walls, perhaps the

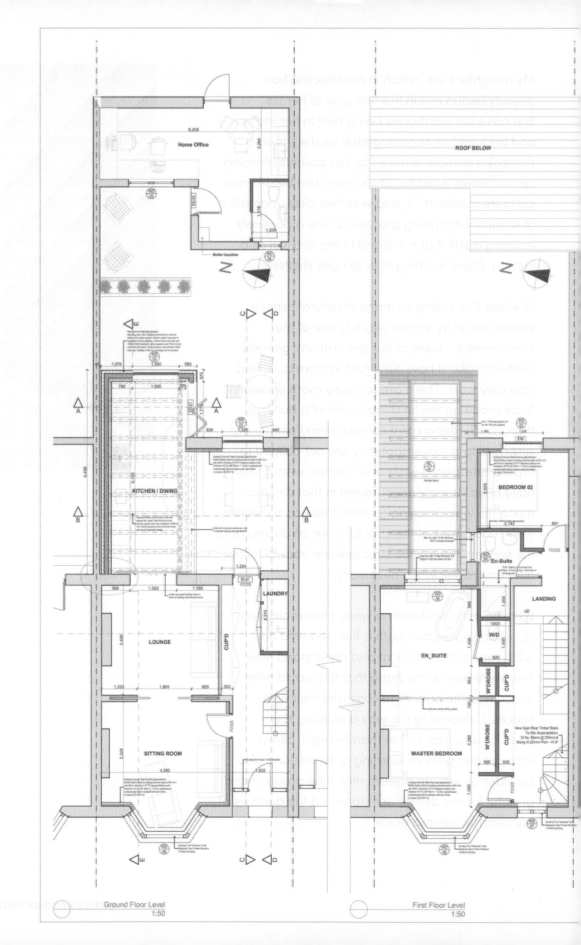

A typical set of construction drawings.

Section AA
1:20

Dining

Exposed Timber Rafters
(No Ridge Beam)

1,730
1,630
1,550
134
4,000
2,758
3,475
2,752
2,400

12 PRINCE ARTHUR TERRACE

01   Proposed Front Elevation
     1:100

12 PRINCE ARTHUR TERRACE

Proposed new
dormer window

2,650   335

Proposed Rear Elevation
1:100

Proposed Section - A-03
1:100

Attic Level
1:50

N

BEDROOM 3

Storge

W.C

Storge

Attic
Hatch/Door To
Eaves Storage

1200 High Wall

1500mm
Head height

2000mm
Head height

1500mm
Head height

New Open Rise Timber Stairs
To Attic Accommodation
13 No. Risers @ 200mm &
Going of 225mm Pitch - 41.6°

1200 High Wall

2,850
2,525   325
2,420
990   2,420   716   2,100
4,082

REV	DATE	DESCRIPTION:	APPROVED

JOB DETAILS

Scale @ A1: 1:50, 1:100, 1:20	Drwg no:	Alssued:
Date:	Description:	
Drawn by:	Plans, Elevations & Section (WIP)	
Checked by:		

different types of materials to be used and the location of doors, windows and features of the home. With a floor plan one should be able to determine the location, layout and room divisions within a structure.

A section is a cut-through view of a structure, seen from one side. Like an orange cut in half, a section will illustrate the vertical make-up of a building, including foundations, flooring, wall heights and roof details. The more complex a structure, the more additional sections and plan views will be needed to illustrate all the different elements. Observing floor plans and sections together allows us to determine the make-up and materials of our structure.

Take some time to explore the floor plan and section of a building on pages 248 and 249. Try to follow every guide available. There is an incredible amount of information to work with.

A disconnected system

It could be said that the ability to measure the strength of modern-day manufactured materials, as well as the predictable loads that will be placed upon them, has created a disconnect between professions. The architect comes up with a design, an engineer has to figure out how it works and how to make it safe, and finally the builder has to make it a reality. Perhaps the recent separation of these professions came about to allow all the individuals involved to focus on their own strength and specialise in their field. I can understand the need for specialists, but I believe that, unfortunately, this separation may have contributed to the breakdown of the home-building process.

The order of operations used to be that we built our homes with the support of our communities, with whatever materials we had available to us. The standard model for home building now requires that an architect designs our house for us, and an engineer decides what materials we should use, often without any consideration of the availability or cost of those materials. Today's builders are often only required to follow the instructions and calculations of book-smart engineers to complete a building, instead of being involved with, and consulting on, the design process, as was standard practice in the past.

This disconnected system can be disempowering for all individuals involved, most of all for the homeowner, as everyone has to rely on the expertise of someone else. When things go wrong, or unsuitable advice is given, no one is willing to take responsibility, and sadly, it is often the client who has to carry the weight of such errors, as they have to live with the implications.

The eleven-step process

1 Organise any necessary paperwork

Understanding your options, obligations and duties as a project manager might sound like the least exciting part of the building process, but it is an important one. Depending on the size and scale of your project, you may be obliged to obtain planning permission and permits for beginning construction work, installing new septic systems and percolation areas. Before you start anything else, it is advisable to get your paperwork in order.

ARCHITECTURAL DRAWINGS

Architectural drawings will be required for planning application. While these drawings do not necessarily have to be completed by an architect, they do have to meet precise guidelines with regard to their scale and detail, so unless you feel confident in this task you may choose to outsource the job.

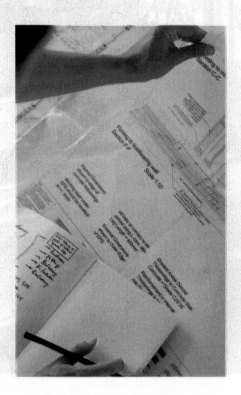

2 Prepare the site

CONSTRUCTION DRAWINGS

These are required for your contractors to know what they need to build and install and where, and also to get your final engineer sign-off on your building. They are typically carried out by an engineer.

PLANNING PERMISSION — IF APPLICABLE

This will involve submitting plans to your local council, completing and submitting soil and percolation tests and giving your neighbours an opportunity to raise any objections.

CLEAR THE SITE

Once you have permission to begin your project you should clear and level the site you intend to build on. This process may require renting a digger or hiring someone to drive a digger for you. Make sure you have a plan for what you will do with the cleared topsoil. You may wish to keep it to one side for redistribution and landscaping at a later point. Just ensure it is well out of the way of your building.

DIGGING

Consult your plans and try to complete all the necessary digging work in one go. Consider your foundations, septic tank, percolation area, land drainage and garden landscaping.

3 Lay foundations

Whether you are using concrete strip foundations, screw pilings, rammed earth tyres or stone, foundations are always where we start a building. This might be the heaviest work of your whole build, so it can be a good time to call in family and friends. Remember to run any underground services such as sewerage or water pipes that need to be passed through the foundations. Refer to pages 32–3 for foundation options and methods.

4 The skeleton framing

Whether you are using timber, concrete block or straw bales to frame your walls, the next step is to construct the walls all the way up to the roof line. Remember to leave openings for doors and windows, and use the appropriate lintels to span over these openings (consult an engineer or an experienced builder on the lintels required if you feel unsure). Once the walls are complete, move straight on to framing the roof. Now the skeleton is complete. If you are building with brick or block you will install your cavity insulation at this point. Refer to pages 138–40 for how to build a timber frame wall.

5 Weatherproofing and doors and windows

The next step is to wrap your building in its breathable membrane or weatherproofing layer. This will protect the work you have done and provide you with a dry or shaded site to work within. Doors and windows can now also be installed so that your external and internal finishes can meet them. Congratulations – you have a weatherproof structure! This is a good moment to take a short break and breathe a little sigh of relief as you go out for dinner knowing that your building is secure. Refer to pages 192–4 on how to correctly weatherproof your building.

6 First fix services

Next up is usually the first installation of services. Heating and cooling systems, as well as your electrical and plumbing, can all be run on or through your walls and throughout the house. The backing boxes for your electrical switches and sockets can also be installed. Most of the time, everything that happens in the first fix will end up being hidden, either in the floor or ceiling cavities or within the walls.

7 Exterior shell

At the same time as the first fix installation, you can be working on the external shell. Timber cladding, plasterwork, roofing metal or tiles, and any other external shell components can now be brought to the finished stage. Any delicate and/or moisture-sensitive materials, such as plasterboard, insulation and flooring, can be brought to site and safely stored in your dry workspace. Refer to pages 205–10 for exterior cladding and roofing options.

8 Installing insulation

At this stage, the insulation can be installed throughout the house, whether it is sheep wool, blown-in cellulose or mineral wool batts. All wall, floor and ceiling cavities can now be filled with your insulation. Refer to pages 174–6 for how to install simple mineral wool insulation.

9 Interior surfaces and cabinetry

This is an exciting stage, as it is now time to dress the interior of your home. Plasterboard, wood panelling or clay and lime renders can all go on. Kitchen cabinets and dressers, and any waterproof backing boards for tiles and shower areas, can be installed, as well as your first coat of paint. All remaining internal finishes can go in, including door and window sills and trims, countertops, tiles and flooring.

10 Second fix

The remaining plumbing and electrical work can now be done. If you are running exposed conduits for your electrics or any exposed plumbing pipework, this is when you will be installing these elements. Taps and plumbing fixtures such as toilets, sinks and bathtubs can all be mounted and connected. All sockets, switches and light fittings can be connected, and the electrical circuit board can be finished. Additional coats of paint can be applied, along with any other final decorative flourishes that you want.

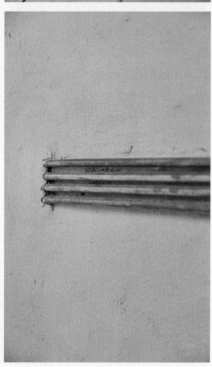

11 Clean-up

The final job is the clean-up. This applies inside and outside your building. Driveways, footpaths and landscaping can all be done in the knowledge that there will be no more heavy machinery or building crews churning up the ground.

Your build is complete. Kettle on, feet up, it's time to enjoy the nest you have built. Once you start to inhabit the space you will undoubtedly find things you would like to change, or ways that the house could suit you better. The good news is that by this point you will have gained enough skill and knowledge that these tasks shouldn't intimidate you. They are simply part of our ever-evolving homes.

ACKNOWLEDGEMENTS

Without the full support and encouragement of my community this book would never have come into being.

Erin, I can never thank you enough for the countless nights you spent learning the content of this book so that you could then help to bring clarity to the pages and ensure that others can learn from them too. Thank you for helping this book reach its potential and be something we both can be incredibly proud of. You are my greatest inspiration.

Thank you, Shantanu, for your patience and your enthusiasm in bringing this idea to life visually and for enabling it to be as beautiful as it could be. We are so grateful for your creative input.

Thank you, Manchán, for believing that this could be more than a PDF hand-out long before anyone else even knew it existed, and for always steering me in the direction of the greater good.

Thank you, Ankur, for reading, re-reading and reading again the many iterations of this book, and for all your help in expanding and refining the story.

Thank you, Joe, for liaising with us from across the world and for interpreting my ever-vague visual requests. You're a great friend and as good a mind reader as they come.

Thank you to Mike Reynolds for giving me the opportunity to teach all those years ago and thus awakening a joy in me I never knew was there. Your upstream approach to life and building will always be an inspiration to me.

Thank you to Marianne Gunn O'Connor for finding a home for this book and for believing in the importance of this story before it even existed.

Thank you to everyone at Gill for creating such a collaborative, shared process in bringing this book to fruition. I am so grateful to be working with you and feel extremely proud of the finished result.

Lastly, thank you to my parents, Ceresse and Tim. Whether it was your intention or not, growing up on a building site was the greatest inheritance you could have given me. Thank you for sharing your many building projects with me and for not shielding me from them. I'm sorry I wasn't nearly as interested then as I would be now. Mum, thank you for teaching me to never shy away from learning something new and to expect adventure around every corner. Dad, thank you for all your support and for being part of my new life here in Ireland. I'm finally ready to build that sailing boat with you.

Index

Notes